Springer Series in Optical Sciences   Volume 65

Editor: Anthony E. Siegman

W0245728

Springer Series in Optical Sciences

Editorial Board: A. L. Schawlow K. Shimoda A. E. Siegman T. Tamir

Managing Editor: H. K. V. Lotsch

Volumes 1–41 are listed on the back inside cover

Francisco J. Duarte (Ed.)

High-Power Dye Lasers

With 93 Figures

Springer-Verlag Berlin Heidelberg GmbH

Dr. Francisco J. Duarte

Photographic Research Laboratories, Photographic Products Group,
Eastman Kodak Company, Rochester, NY 14650-1744, USA

ISBN 978-3-662-13813-7   ISBN 978-3-540-47385-5 (eBook)
DOI 10.1007/978-3-540-47385-5

Typesetting: Macmillan, India
Production Editor: P. Treiber

54/3140 – 5 4 3 2 1 0 – Printed on acid-free paper

Preface

Pulsed dye lasers are a formidable source of high-power coherent tunable radiation in the visible part of the spectrum. In addition, the versatile nature of these lasers ensures their applicability to a wide range of scientific and industrial tasks.

The aim of this book is to integrate a number of topics relevant to the generation of high-power tunable dye laser radiation. These elements of high-power dye lasers are considered in chapters devoted to dispersive dye lasers, gain analysis and amplifier design, excimer-laser pumped dye lasers, copper-vapor-laser pumped dye lasers, and flashlamp pumped dye lasers. A characteristic of the presentations is their technological perspective, in addition to considering the relevant physics. As such, this book should be of interest to all those familiar with laser physics and optics who are working on dye lasers or applying dye lasers.

Furthermore, it is our hope that many of the topics considered here will offer a useful extension of Schäfer's pioneering book *Dye Lasers* (Topics Appl. Phys., Vol. 1).

For useful discussions throughout the duration of this project I am indebted to D. R. Foster, J. J. Ehrlich, L.W. Hillman, A. E. Siegman, and the contributing authors.

Rochester, November 1991 *F. J. Duarte*

Contents

Contributors

Duarte, Francisco J.
Photographic Research Laboratories, Photographic Products Group,
Eastman Kodak Company, Rochester, NY 14650-1744, USA

Everett, Patrick N.
Massachusetts Institute of Technology, Lincoln Laboratory, L-203,
Lexington, MA 02173-9108, USA

Jensen, Craig C.
Los Alamos National Laboratory, P.O. Box 1663, J564, Los Alamos,
NM 87545, USA

Tallman, Charles R.
Los Alamos National Laboratory, P.O. Box 1663, E543, Los Alamos,
NM 87545, USA

Tennant, Roger A.
Los Alamos National Laboratory, P.O. Box 1663, J565, Los Alamos,
NM 87545, USA

Webb, Colin E.
Atomic and Laser Physics, Department of Physics, University of Oxford,
The Clarendon Laboratory, Parks Road, Oxford, OX1 3PU, UK

1. Introduction

Francisco J. Duarte

High-power tunable laser radiation is desirable for numerous applications. Dye lasers have been shown to yield high-power tunable coherent radiation both in the single-pulse regime and in the high-pulse repetition frequency (prf) domain.

In this book we present and discuss in detail physics and technology elements of high-power pulsed tunable dye lasers.

In this brief introduction the discussion is limited to some generalized aspects relevant to high-power dye lasers and to providing a perspective of the topics selected for discussion.

Firstly, it is important to define the meaning of *high power*. From our perspective, high power refers to high *average* power. That is, the elements considered in this book provide the physics and technological background of concepts and components that are utilized in, or could be integrated into, dye laser systems capable of delivering average powers of 100 W or more. With regard to this, the scope of the book is inherently limited to pulsed systems. However, there is a great degree of flexibility in the regimes of operation that may offer such levels of output power. On the one hand, one could consider a dye laser system capable of delivering a few hundred Joules per pulse at a relatively low prf, as may be typical of a flashlamp-pumped dye laser system. On the other, the alternative of relatively low-pulse energies at a prf exceeding 10 kHz is available from dye laser systems excited by metal vapor lasers.

At this stage it may be worth mentioning an issue of fundamental importance that is relevant to *all* high-average power laser systems, that is, the topic of *heat dissipation*. In the area of heat removal, dye solutions enable fast flow speeds ($5-10 \, \mathrm{m \, s^{-1}}$ or more) that are essential to the process of heat removal. Hence, dye lasers are capable of operation in the kHz regime and can deliver high-average powers.

A perspective on the status of high-power tunable dye lasers is provided in Table 1.1. The listing includes information on high-pulse energy and high average-power dye lasers. From the table it is clearly evident that dye lasers are a formidable source of powerful tunable laser radiation in the visible.

Some comments on the figures quoted in this table may offer further elucidation. The high-pulse energy in the flashlamp-pumped dye laser reported by *Baltakov* et al. [1.1] was obtained using coaxial excitation at a conversion efficiency of 0.8%. The pulse length was about 10 μs and the peak power was reported to be 40 MW. The emission bandwidth is quoted at 9 nm (full width) [1.1]. The relatively high prf device discussed by *Mazzinghi* et al. [1.2] has

Springer Series in Optical Sciences, Vol. 65
High-Power Dye Lasers   Editor: Francisco J. Duarte
© Springer-Verlag Berlin Heidelberg 1991

Table 1.1. High Power Dye Lasers

Excitation source	prf	Pulse energy	Average power	Dye	Reference
Flashlamp	Low	400 J		Rhodamine 6G	[1.1]
Flashlamp	50 Hz	~ 5 J	200 W	Rhodamine 6G	[1.2]
Flashlamp[a]	850 Hz	1.4 J	~ 1.2 kW	Coumarin 504	[1.3]
Flashlamp[a]	10 Hz	140 J	~ 1.4 kW	Rhodamine 6G	[1.4]
Excimer laser	Very Low	800 J		Coumarin 480	[1.5]
CVL[b]	≥ 10 kHz		> 2 kW	Rhodamine	[1.6]

[a] Burst mode.
[b] Copper-vapor laser.

a waveguide excitation configuration that provides a conversion efficiency of up to 0.6%. Certainly the energy per pulse at high prf is lower than the energy measured at low prf quoted in Table 1.1. This flashlamp-pumped planar waveguide dye laser was also operated in a relatively narrow-linewidth regime using an open cavity grazing-incidence configuration. The reported linewidth was 10 GHz and the energy loss in this regime of operation was some 20% [1.2]. The high prf device reported by *Morton* and *Draggoo* [1.3] employed dye flow speeds in the 7–10 m s^{-1} range. The laser pulse length was 1.5 μs (FWHM) and the reported average power was in a five second burst [1.3]. This device represents one of the early tunable laser systems developed for uranium enrichment. An additional high-average power flashlamp-pumped dye laser is that developed under the dye laser program of the U.S. Army (Missile Command) [1.4] at the Avco Research Laboratory. This device yields a high-energy pulse (5 μs at FWHM) at a relatively low prf. The measured conversion efficiency was 1.8% [1.4]. Improvements in energy conversion efficiency in this class of laser are, in part, due to the use of converter dyes [1.7].

Again, in the area of laser excited dye lasers there is the option of high pulse energies or high average powers. In the work of *Tang* et al. [1.5] a high-energy excimer laser-pumped dye laser is discussed. In this experiment a large-scale XeCl laser was employed to excite a volume of coumarin 480 dye. The reported 800 J were delivered in a 500 ns pulse at a photon conversion efficiency of 27% [1.5].

The efficiency of copper-vapor lasers can be in the 1–3% range [1.8] and the photon conversion efficiency from the excitation laser to the dye laser may approach 50% [1.9]; see also Chaps. 2 and 5. The high-prf copper-vapor-laser-pumped dye laser system at Lawrence Livermore National Laboratory [1.6] employs multiple stages of amplification. The impressive feature of this high-average power (in excess of 2 kW [1.6]) is that it is in the form of highly coherent (single frequency) tunable radiation that originates at the master oscillator.

Certainly, this is a powerful, efficient, and highly reliable tunable laser system that exploits well the physics of dye lasers.

From a speculative perspective it may be stated that the ideal tunable laser, for high average-power operation, could be a gas-phase discharge-excited dye laser. Possible advantages may include high efficiencies, due to direct excitation [1.10, 11], high active medium homogeneity [1.12] and volume scalability. Although optically pumped gas-phase dye lasers have been known for some time [1.13–15], laser action resulting from direct discharge excitation remains to be demonstrated.

1.1 Objectives

The main objective of this work is to provide an up-to-date and detailed survey of the physics and technological issues relevant to high-power dye lasers. As such, the scope is limited to elements of pulsed dye lasers applicable to high power generation and to components that can be integrated into high-power dye laser systems. These elements of pulsed dye lasers are represented by a selection of topics ranging from the basic and general to the specialized and specific.

Furthermore, it is hoped that the material included and the style used by the contributors will be suitable for readers searching for useful and practical information.

Given the scope of the present work, several familiar and important topics of dye lasers are not considered, or treated only in a peripheral sense, and the reader is referred to recognized sources of information on these subjects [1.16–18]. Here, we refer specifically to topics such as historical perspectives, continuous wave dye lasers, femtosecond dye lasers, laser dyes, and dye laser applications.

An additional objective of this work is to serve as a complementary volume to the pioneering work of *Schäfer* [1.16]. With regard to this, the topics considered in this book have little overlap with most of the subjects discussed in [1.16], and thus offer a valuable extension. A similar comment can be made in reference to the work of *Duarte* and *Hillman* [1.17].

In general, the contributions presented in this book may be more suitable for those already familiar with the disciplines of lasers and optics. For background information, readers may refer to [1.16, 17, 19].

1.2 Brief Survey of Contents

In addition to this short introduction, the present work includes five chapters on different aspects of high-power dye lasers. The next two chapters, on dispersive dye lasers and gain analysis, cover subject material of a basic nature and are thus relevant to the subsequent chapters. The discussion then centers on excimer-laser-pumped dye lasers and continues with a description of high-prf copper-vapor-laser-pumped dye lasers. The concluding chapter considers flashlamp-pumped dye lasers in detail.

Chapter 2, on dispersive dye lasers, was written to provide a useful introduction to the subject of narrow-linewidth dispersive dye lasers. As such, the main emphasis revolves around the design and technology aspects of dispersive dye laser oscillators. With regard to this, detailed design criteria on multiple-prism grating assemblies are discussed using examples. The necessary theory is outlined in a brief introductory section without details of derivation. The background of the theoretical section is given by *Duarte* [1.20] and references therein. A section on multiple-slit interference is introduced as an application of both dye lasers and beam propagation through a dispersive optical system. An appendix relevant to pulse compression in femtosecond prismatic dye lasers is also included.

Chapter 3 considers laser gain and amplifier design. As such, this chapter provides a thorough description of the dynamics of dye laser excitation. Particular attention is given to excimer laser excitation. The discussion then considers fluorescence yield, dye transmission and gain. A specific section is devoted to quantum yield and the chapter concludes with a detailed discussion of dye laser amplifiers.

Chapter 4 is dedicated to large-scale excimer-laser-pumped dye lasers. This chapter describes numerous engineering and technological issues associated with this class of laser. The authors consider dye-cell design, flow requirements, and issues related to dyes and solvents. The topic of optical damage, evident in dye cells, associated with a high energy flux is also considered. The discussion also includes topics relevant to the technology of the excimer laser. This work concludes with an assessment of the performance of a specific excimer-laser-pumped dye laser.

Chapter 5 refers to high-prf copper-vapor-laser-pumped dye lasers. This contribution starts with a description of elemental copper vapor lasers and copper halide lasers An entire section is then devoted to copper vapor laser beam control. This discussion includes plane–plane cavities, unstable resonators, injection controlled oscillators and amplifiers. A theoretical description of copper-vapor-laser-pumped dye lasers focuses on gain, saturation intensity, and pumping under longitudinal and transverse geometries. The author then considers the performance of broadband dye laser oscillators, narrow-linewidth dye laser oscillators, and oscillator-amplifiers. The chapter concludes with a description of important areas of application for copper-vapor-laser-pumped dye lasers.

Chapter 6 provides a detailed description of flashlamp-pumped dye lasers. Here, the reader is first introduced to the technology of flashlamps, then the spectroscopic characteristics of the dye are considered, followed by a discussion of rate equations. The author next considers efficiency improvement using spectral conversion and describes flow systems and solvent recycling. The subject of resonators and beam propagation in flashlamp-pumped dye lasers is discussed with attention given to unstable resonators and waveguide resonators. The chapter concludes with a discussion on spectral control. An appendix indicates how the notation in Chap. 6 corresponds to that of Chap. 3.

1.3 Perspective

It is clearly evident that dye lasers have provided a major contribution to the basic development and general understanding of laser physics. As examples, we may cite two areas of interest. Firstly, the introduction of useful and highly practical cavity configurations for narrow-linewidth emission (see, for example, [1.20] and references therein) that have been subsequently applied to gas and solid state lasers [1.20]. Moreover, the study of laser dynamic instabilities has gained significantly from dye laser research; see, for example, [1.21–25].

From a technological perspective, dye lasers offer a formidable source of coherent tunable radiation in the visible portion of the spectrum. This is particularly true in the area of high-average power generation. From a more general point of view, it should be stated that dye lasers have demonstrated unparalleled operational flexibility: from continuous-wave oscillation to high energy pulse emission and from long pulses to femtosecond pulse generation.

As indicated previously [1.17], it is possible, by careful consideration of the physics and technology, to design highly reliable and efficient dye laser systems where the problems of dye lifetime are overcome.

At this stage it should be reiterated that the strength of dye lasers is in the visible spectrum. In the near infrared, new tunable solid state sources offer a very effective and attractive alternative. This is mainly due to the fact that many infrared dyes are somewhat inefficient but more importantly these dyes need solvents that may be classified as toxic. This appears to be an issue of basic importance for the future development and success of dye lasers in this region of the spectrum. Hence, in direct reference to dye molecular engineering, solubility is a very important parameter. New dyes should be highly soluble in solvents such as water, ethanol, methanol, or a combination of these. Recent progress in molecular development has been reported by *Pavlopoulos* et al. [1.26] and *Chen* et al. [1.27]. For instance, some of the coumarin tetramethyl compounds reported in [1.27] have been found not only to be more efficient than the parent compound, but are freely soluble in ethanol to concentrations approaching 10^{-1} M and have been demonstrated to lase at these high number densities under excimer-laser excitation.

Dye lasers have already demonstrated average powers in the kW regime. Further scaling does not appear to be limited by factors of a fundamental nature.

References

1.1   F.N. Baltakov, B.A. Barikhin, L.V. Sukhanov: JETP Lett. **19**, 174 (1974)
1.2   P. Mazzinghi, P. Burlamacchi, M. Matera, H.F. Ranea-Sandoval, R. Salimbeni, U. Vanni: IEEE J. Quantum Electron. **QE-17**, 2245 (1981)
1.3   R.G. Morton, V.G. Draggoo: IEEE J. Quantum Electron. **QE-17**, (12), 222 (1981)
1.4   J.J. Ehrlich: Private communication
1.5   K.Y. Tang, T. O'Keefe, B. Treacy, L. Rottler, C. White: Kilojoule output XeCl dye laser: optimization and analysis, in *Proceedings: Dye Laser/Laser Dye Technical Exchange Meeting*, ed. by J.H. Bentley (U.S. Army Missile Command, Redstone Arsenal 1987) pp. 490–502
1.6   J.A. Paisner: Private communication
1.7   P.N. Everett, H.R. Aldag, J.J. Ehrlich, G.S. Janes, D.E. Klimek, F.M. Landers, D.P. Pacheco: Appl. Opt. **25**, 2142 (1986)
1.8   A.Y. Artemev, B.L. Borovich, L.A. Vasilev, V.E. Gerts, E.P. Nalegach, S.A. Negashev, E.G. Radostin, V.M. Rybin, V.M. Ryazanskii, L.V. Tatarintsev, A.N. Ulyanov: Sov. J. Quantum Electron. **12**, 456 (1982)
1.9   W.W. Morey: Copper vapor laser pumped dye laser, in Proceedings of the International Conference on Lasers '79, ed. by V.J. Corcoran (STS Press, McLean, VA 1980) pp. 365–373
1.10 P.W. Smith, P.F. Liao, C.V. Shank, T.K. Gustafson, C. Lin, P.J. Maloney: Appl. Phys. Lett. **25**, 144 (1974)
1.11 P.W. Smith, P.F. Liao, C.V. Shank, C. Lin, P.J. Maloney: IEEE J. Quantum Electron. **QE-11**, 84 (1975)
1.12 N.A. Borisevich: Opt. Acta: **32**, 1071 (1985)
1.13 N.A. Borisevich, I.I. Kalosha, V.A. Tolkachev: Zh. Prikl. Spektrosk. **9**, 1108 (1973)
1.14 B. Steyer, F.P. Schäfer: Opt. Commun. **10**, 219 (1974)
1.15 N.G. Basov, O.A. Logunov, D.K. Nurligareev, K.K. Trusov: Sov. J. Quantum Electron. **11** 1400 (1981)
1.16 F.P. Schäfer (ed.): *Dye Lasers*, Topics Appl. Phys., Vol. 1, 3rd edn. (Springer, Berlin, Heidelberg 1990)
1.17 F.J. Duarte, L.W. Hillman (eds.): *Dye Laser Principles* (Academic, New York 1990)
1.18 M. Maeda: *Laser Dyes* (Academic, New York 1984)
1.19 A.E. Siegman: *Lasers* (University Science Books, Mill Valley, CA 1986)
1.20 F.J. Duarte: Narrow-linewidth pulsed dye laser oscillators, in *Dye Laser Principles*, ed. by F.J. Duarte, L.W. Hillman (Academic, New York 1990) pp. 133–183
1.21 L.W. Hillman, J. Krasinski, R.W. Boyd, C.R. Stroud: Phys. Rev. Lett. **52**, 1605 (1984)
1.22 L.W. Hillman, J. Krasinski, K. Koch, C.R. Stroud: J. Opt. Soc. Am. B **2**, 211 (1985)
1.23 F.J. Duarte, J.A. Piper: Appl. Opt. **23**, 1391 (1984)
1.24 F.J. Duarte, J.J. Ehrlich, S.P. Patterson, S.D. Russell, J.E. Adams: Appl. Opt. **27**, 843 (1988)
1.25 H. Fu, H. Haken: Phys. Rev. A **42**, 4151 (1990)
1.26 T.G. Pavlopoulos, J.H. Boyer, M. Shah, K. Thangaraj, M. Soong: Appl. Opt. **29**, 3885 (1990)
1.27 C.H. Chen, J.L. Fox, F.J. Duarte, J.J. Ehrlich: Appl. Opt. **27**, 443 (1988)

2. Dispersive Dye Lasers

Francisco J. Duarte

With 10 Figures

In this chapter we provide a succinct introduction to the elements and architecture of dispersive dye lasers. The emphasis is on the technology and the intention is to provide a useful guide to the subject for physicists and engineers already familiar with the basic elements of laser theory and operation. With regard to this, the relevant formulae are presented without derivation or discussion except when judged necessary. Original references are given to provide further details to those interested in the physics and derivation of the formulae.

The focus is on the design and construction of efficient dispersive dye laser oscillators that could be integrated in larger high-power oscillator-amplifier systems and high-power master-oscillator forced-oscillator systems.

The initial discussion is focused on generalized cavity concepts and on the integration of various cavity optical components in the architecture of basic resonators. Appropriate mathematical formulae needed to design and evaluate the performance of the various resonators are then given in a theoretical section. The description continues with a discussion of the various mechanical and physical requirements on the architecture needed to optimize laser performance. In order to illustrate the use of mathematical formulae and the design principles outlined, specific design examples are considered.

For a historical perspective on this subject and further technical details, readers should refer to [2.1, 2]. Excitation sources are reviewed by *Duarte* [2.3]. For detailed descriptions of excimer lasers, refer to Chap. 4. The copper-vapor laser (CVL) is considered in Chap. 5 and flashlamps are reviewed in Chap. 6.

2.1 Resonator Outlines

In general, dye laser resonators can be classified in two large categories: those that function with their intrinsic narrow beam waist and those that utilize intracavity beam expansion. The former class includes cavities that achieve linewidth narrowing using Littrow gratings in conjunction with intracavity etalons and pure grazing-incidence resonators with and without intracavity etalons. The latter class can be divided into two subclasses: resonators involving two-dimensional telescopic beam expansion and oscillators utilizing one-dimensional prismatic beam expansion. In this class of cavity, the gratings can be

Springer Series in Optical Sciences, Vol. 65
High-Power Dye Lasers Editor: Francisco J. Duarte
© Springer-Verlag Berlin Heidelberg 1991

either in Littrow or near grazing-incidence configuration and the use of intracavity etalons is optional.

In this section, we describe the architecture of these resonators and their
basic characteristics. All the resonators considered here are of the closed cavity
type, that is, configurations where the output beam exits the cavity via an output
coupler mirror rather than from a reflection loss of an intracavity element or
a grating [2.2].

2.1.1 Class I: Intrinsic Narrow Beam Waist Resonators

This is perhaps the most elementary of the wavelength selective dye laser
resonators. The basic design utilizes a simple grating-mirror cavity [2.4]. In this
resonator, wavelength tunability and linewidth narrowing is provided by the
grating in Littrow mounting. Further tuning and linewidth narrowing options
are provided by the use of one or more intracavity etalons [2.5–7] as illustrated
in Fig. 2.1a. This approach has the advantage of being simple and can readily
employ commercially available gratings and etalons. Disadvantages include
a high intracavity energy flux, which may damage the grating, the coatings on
the etalons, and losses introduced by the etalons. Characteristics of this type of
resonator are summarized in Table 2.1.

An additional class of resonator that employs the natural divergence of the
intracavity beam for full illumination of the grating is the grazing-incidence
design [2.8–12]. In these lasers the grating is used at a high angle of incidence
and tuning is achieved by rotating the mirror that reflects the diffracted order
back to the grating. Since this type of configuration allows total illumination of
the grating, the resulting laser linewidths can be very narrow. Indeed, single-
longitudinal-mode operation was reported by *Littman* [2.10, 11]. A variation of
this configuration involves the use of a second grating in Littrow configuration
rather than the tuning mirror [2.10]. Use of intracavity etalons has also been

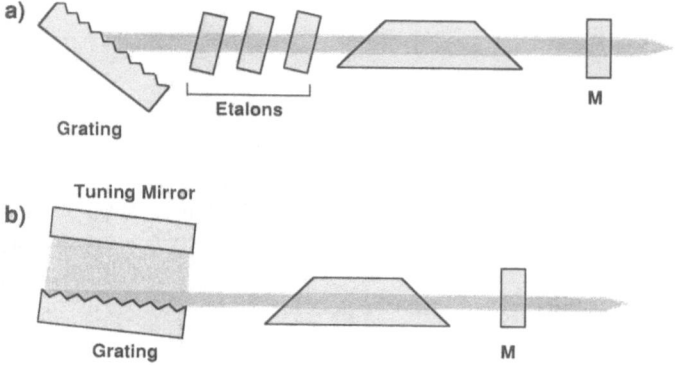

Fig. 2.1a,b. Intrinsic narrow-beam waist resonators. (**a**) Grating-mirror cavity incorporating intracavity etalons. (**b**) Grazing-incidence grating design

Table 2.1. Characteristics of narrow beam waist resonators

Cavity configuration	Cavity length[a] [cm]	Polarization	Alignment	Remarks
Littrow grating	5–10	Depends on grating[b]	Very simple	Efficiency[a] can be ⩾ 20%
Littrow grating and etalon	~ 10	Depends on grating[b]	Simple	
Multiple etalons	10–15	Unpolarized	Simple	$\Delta v \sim 4\,\text{GHz}$ [2.7]
Grazing incidence	5–10	Depends on grating[b]	Simple	$150\,\text{MHz} \leqslant \Delta v \leqslant 2.5\,\text{GHz}$ [2.9–11] and effs.[a] vary from 2–4%

[a] Refers to laser-pumped dye lasers exclusively.
[b] See refs. in Sect. 2.2.3 for polarization characteristics of gratings.

reported to yield single-longitudinal-mode oscillation [2.12]. The advantage of these cavities is their compactness, which helps to achieve single-mode operation. Since narrow-linewidth oscillation can be achieved without the use of intracavity elements, these cavities offer simplicity and their dispersive properties depend solely on the grating. The main disadvantage results from the low efficiency of the grating when used at high angles of incidence, thus limiting the laser conversion efficiency to less than a few percent. These cavities are illustrated in Fig. 2.1b and their characteristics are summarized in Table 2.1.

2.1.2 Class II: Resonators Utilizing Intracavity Beam Expansion

A key element in achieving narrow-linewidth oscillation is the enhancement of the intracavity dispersion available from the grating, which is the main tuning element. An alternative method to illuminate the whole grating, in the more efficient Littrow and near grazing-incidence configurations, is the use of intracavity beam expansion as illustrated in Fig. 2.2.

Two-dimensional intracavity beam magnification is achieved using beam expanders such as transmission or reflective telescopes. Intracavity beam magnification utilizing transmission telescopes [2.13] is illustrated in Fig. 2.3a. These telescopes can be of the Galilean or Newtonian type. The use of the Cassegrian telescopes [2.14, 15] is illustrated in Fig. 2.3b. Cavities employing two-dimensional beam expansion vastly reduce the risk of damage to the grating and offer good efficiencies. This characteristic is true for all resonators incorporating intracavity beam expansion. For further linewidth reduction, an intracavity etalon can be inserted between the telescope and the grating [2.13]. Among the disadvantages of resonators using two-dimensional beam expansion is the

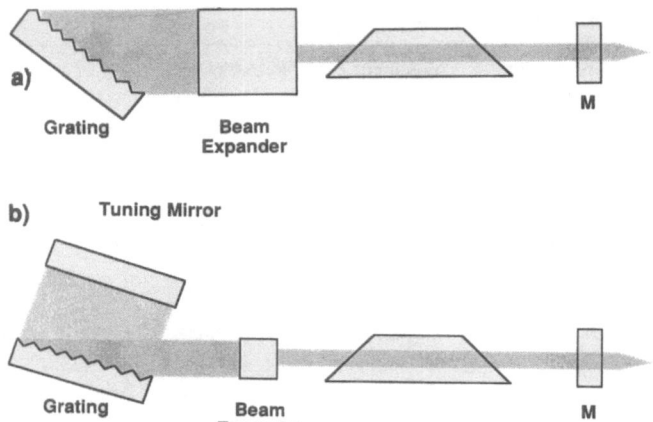

Fig. 2.2a, b. Resonators utilizing intracavity beam expansion. (a) Intracavity beam expansion utilized in conjunction with a grating in Littrow configuration and (b) intracavity beam expansion used in conjunction with a grating in near grazing-incidence configuration

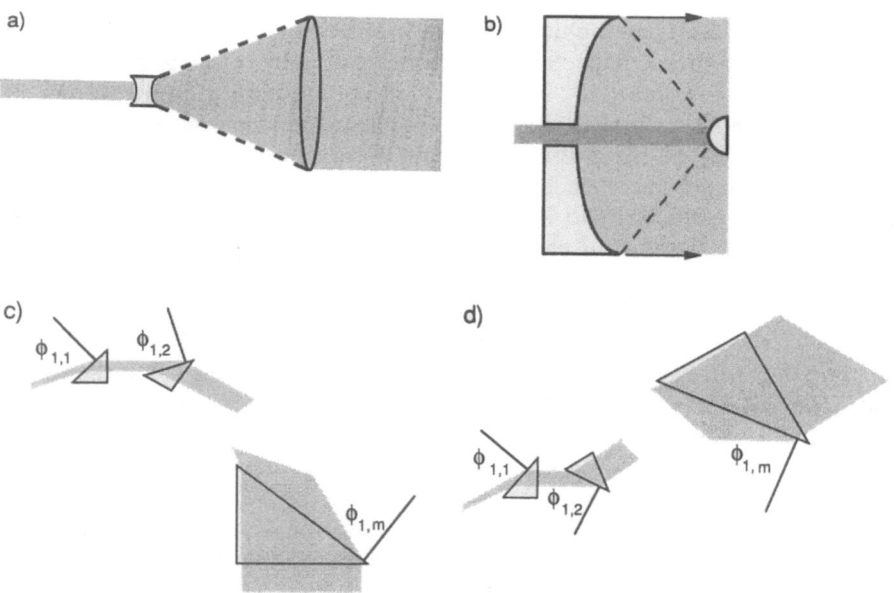

Fig. 2.3. Two-dimensional beam expansion using (a) Galilean telescope and (b) Cassegrain telescope. One-dimensional beam expansion utilizing an additive multiple-prism configuration (c) and a compensating multiple-prism configuration (d)

Table 2.2. Characteristics of resonators utilizing intracavity beam expansion

Cavity configuration	Cavity length[a] [cm]	Polarization	Alignment	Remarks
Littrow grating and telescope	20–35	Depends on grating[c, d]	Difficult	300 MHz[b] ≤ Δv ≤ 2.5 GHz and effs.[a] are in the 2[b]–20% range [2.13]
Littrow grating and Cassegrain telescope[e]	20–30	Depends on grating[c]	Difficult	Δv ∼ 9 GHz at an eff. of 16% [2.15]
MPL[g]	10–15	∼ 100% p-polarized	Easy	60 MHz[f] ≤ Δv ≤ 1.61 GHz [2.20–23] and effs.[a] are in the 5–15% range
HMPGI[g]	10–15	∼ 100% p-polarized	Easy	138 MHz ≤ Δv ≤ 1.15 GHz [2.22–25] and effs.[a] are in the 4–10% range

[a] Refers to laser-pumped dye lasers exclusively.
[b] Reported figure obtained with intracavity etalon.
[c] See refs. in Sect. 2.2.3 for polarization characteristics of gratings.
[d] The telescopic oscillator reported by *Hänsch* [2.13] incorporated in an intracavity Glan-Thompson polarizer.
[e] All figures given in [2.15].
[f] Linewidth obtained with intracavity etalon [2.21].
[g] Linewidth values reported for long-pulse flashlamp-pumped MPL and HMPGI oscillators are in the 138 MHz ≤ Δv ≤ 375 MHz range [2.23, 25].

requirement of a large circular grating and a relatively difficult alignment. Also, these telescopes add significantly to the overall length of the cavity, thus making these configurations less compact. Reported performance for these telescopic resonators is listed in Table 2.2.

An alternative to two-dimensional telescopic beam expansion is the one-dimensional prismatic beam magnification [2.16–23]. As indicated in the literature [2.20] for any given beam magnification value (M) multiple-prism beam expansion is considerably more efficient than single-prism beam expansion. Thus, in this chapter we consider intracavity multiple-prism beam expansion exclusively. Multiple-prism beam expansion is illustrated in Fig. 2.3c and 2.3d. Narrow-linewidth dye lasers incorporating multiple-prism beam expanders deploy their gratings either in Littrow or near grazing-incidence configuration. Prismatic laser cavities utilizing Littrow gratings are known as multiple-prism Littrow (MPL) dye laser oscillators, and prismatic resonators employing gratings in a near grazing-incidence configuration are known as hybrid multiple-prism grazing-incidence (HMPGI) dye laser oscillators [2.22–25]. Thus, the use of a multiple-prism beam expander in the basic Littrow configuration shown in Fig. 2.2a results in a MPL oscillator and the utilization of a multiple-prism expander in the near grazing-incidence configuration shown in Fig. 2.2b results in a HMPGI oscillator.

Note that the number of prisms usually varies between two and four and they can be arranged either in additive or compensating configurations [2.25–27]. The advantage of these cavities over the telescopic resonators is two fold; firstly they are more compact and secondly they provide one-dimensional beam expansion, thus improving alignment requirements. Also, they tend to be more stable thermally. Relative to pure grazing-incidence cavities, MPL and HMPGI oscillators can be considerably more efficient.

MPL and HMPGI oscillators appear to have two disadvantages relative to grazing-incidence resonators; they need additional intracavity space to deploy the multiple-prism expander, and the dispersion of the prismatic array may tend to detract from the achromaticity offered by the simpler grazing-incidence resonator. In the case of HMPGI oscillators, beam magnification factors are often less than 30, thus requiring little extra intracavity space since a couple of small prisms can provide that magnification factor rather efficiently. For MPL oscillators beam magnification factors can be as high as 200, thus requiring extra stages of expansion involving larger prisms. Characteristics of these oscillators are included in Table 2.2.

2.2 Theory

In this section a summary of useful equations relevant to the description of the physics and design of dispersive resonators is presented. The equations are given without derivation, except when considered necessary, and appropriate original references are quoted to provide additional information to those interested in the derivation or further details.

2.2.1 Ray Transfer Matrices

We introduce here the various ray matrices useful to describe intracavity beam propagation. It should be noted that the use of this matrix notation assumes a Gaussian beam profile, which may not always be the case. However, experience indicates that lasing in a TM_{00} and a single longitudinal mode provide near-Gaussian beam profiles. In this case, many authors use Gaussian beam propagation characteristics to analyze dye laser resonators [2.9, 13, 28–32]. For a detailed description of $ABCD$ ray matrices the reader should refer to [2.33–35].

The most basic ray matrix of interest to dye laser resonators is the $ABCD$ matrix representing propagation in free space through a distance L [2.34, 35]

$$\begin{pmatrix} A & B \\ C & D \end{pmatrix} = \begin{pmatrix} 1 & L \\ 0 & 1 \end{pmatrix}.$$

(2.1)

For a thin convex (positive) lens of focal length f the propagation matrix is given by [2.34, 35]

$$\begin{pmatrix} A & B \\ C & D \end{pmatrix} = \begin{pmatrix} 1 & 0 \\ -1/f & 1 \end{pmatrix}. \tag{2.2}$$

For a concave (negative) lens of focal length f the matrix has the same format as (2.2) but with $C = 1/|f|$.

Following *Siegman* [2.35] one can write the matrix for a well-adjusted Galilean telescope as

$$\begin{pmatrix} A & B \\ C & D \end{pmatrix} = \begin{pmatrix} f_2/|f_1| & f_2 - |f_1| \\ 0 & |f_1|/f_2 \end{pmatrix}. \tag{2.3}$$

In this notation the beam magnification provided by the telescope is $M = f_2/|f_1|$, where f_2 is the focal length of the convex lens and $|f_1|$ is the focal length of the concave lens. Note that in practical telescopes $f_2 \gg |f_1|$, so that magnification factors of about 50 are routinely provided. The length of the telescope is given by the B term of (2.3). For a well-adjusted Newtonian telescope the transfer matrix can be written as

$$\begin{pmatrix} A & B \\ C & D \end{pmatrix} = \begin{pmatrix} -f_2/f_1 & f_2 + f_1 \\ 0 & -f_1/f_2 \end{pmatrix} \tag{2.4}$$

where $f_2 \gg f_1$.

Using the results given by *Siegman* [2.36] the matrix for a flat grating can be written as

$$\begin{pmatrix} A & B \\ C & D \end{pmatrix} = \begin{pmatrix} \cos\Theta_2/\cos\Theta_1 & 0 \\ 0 & \cos\Theta_1/\cos\Theta_2 \end{pmatrix} \tag{2.5}$$

where Θ_1 is the angle of incidence and Θ_2 is the angle of diffraction. For the important case of a grating used in Littrow configuration where $\Theta_1 = \Theta_2$ this matrix reduces to

$$\begin{pmatrix} A & B \\ C & D \end{pmatrix} = \begin{pmatrix} 1 & 0 \\ 0 & 1 \end{pmatrix}, \tag{2.6}$$

and is the same matrix for a flat mirror used with the angle of incidence equal to the angle of reflection.

The single-pass $ABCD$ transfer matrix for a multiple-prism beam expander composed of r prisms can be written as [2.2, 37]

$$\begin{pmatrix} A & B \\ C & D \end{pmatrix} = \begin{pmatrix} \left[\prod_{m=1}^{r} k_{1,m} \prod_{m=1}^{r} k_{2,m} \right] & B \\ 0 & \left[\prod_{m=1}^{r} k_{1,m} \prod_{m=1}^{r} k_{2,m} \right]^{-1} \end{pmatrix} \tag{2.7}$$

where

$$B = l\left[\left(\prod_{m=1}^{r} k_{1,m}\right)\left(\prod_{m=1}^{r} k_{2,m}\right)^{-1}\sum_{m=1}^{r}(n_m)^{-1}\left(\prod_{j=1}^{m} k_{1,j}\right)^{-2}\left(\prod_{j=m}^{r} k_{2,j}\right)^{2}\right]$$

$$+ D\left[\left(\prod_{m=1}^{r} k_{1,m}\right)\left(\prod_{m=1}^{r} k_{2,m}\right)\sum_{m=1}^{r-1}\left(\prod_{j=1}^{m} k_{1,j}\right)^{-2}\left(\prod_{j=1}^{m} k_{2,j}\right)^{-2}\right].$$

(2.8)

In the derivation of this matrix it is assumed that all prisms are separated by a uniform distance D and that the optical path l through each prism is the same. In (2.7) the beam expansion coefficients at the incidence and exit planes are given by

$$k_{1,m} = \frac{\cos\psi_{1,m}}{\cos\phi_{1,m}}$$

(2.9a)

and

$$k_{2,m} = \frac{\cos\phi_{2,m}}{\cos\psi_{2,m}}$$

(2.9b)

respectively, where $\phi_{1,m}$ and $\phi_{2,m}$ are the incidence and exit angles at the mth prism. The incidence and refraction angles are related by the well-known expression

$$\sin\phi_{1,m} = n(\lambda)\sin\psi_{1,m} .$$

(2.10)

In these equations the overall beam magnification is given by

$$M = \prod_{m=1}^{r} k_{1,m} \prod_{m=1}^{r} k_{2,m} .$$

(2.11)

For a multiple-prism beam expander designed for orthogonal beam exit so that $\phi_{2,1} = \phi_{2,2} = \ldots = \phi_{2,m} = 0$, and $\psi_{2,1} = \psi_{2,2} = \ldots = \psi_{2,m} = 0$, (2.11) reduces to the well-known result

$$M = \prod_{m=1}^{r} k_{1,m}$$

(2.12)

and (2.7) reduces to the expression given by *Duarte* [2.37]. Furthermore, it is easy to show that (2.7) reduces to the result discussed by *Turunen* [2.38] and *Tache* [2.39] for $r = 1$.

For those interested in performing a round-trip stability analysis of a prismatic cavity, the matrix for the return pass is given by

$$
\begin{pmatrix} A & B \\ C & D \end{pmatrix} = \begin{pmatrix} \left[\displaystyle\prod_{m=1}^{r} k_{1,m} \prod_{m=1}^{r} k_{2,m} \right]^{-1} & B \\ 0 & \displaystyle\prod_{m=1}^{r} k_{1,m} \prod_{m=1}^{r} k_{2,m} \end{pmatrix} .
$$
(2.13)

The ray transfer matrix for a slab with parallel surfaces and material with refractive index n can be written as

$$
\begin{pmatrix} A & B \\ C & D \end{pmatrix} = \begin{pmatrix} 1 & (l_e/n)(\cos \phi_e/\cos \psi_e)^2 \\ 0 & 1 \end{pmatrix}
$$
(2.14)

where l_e is the optical path length, ϕ_e the angle of incidence and ψ_e the angle of refraction. For normal incidence (2.14) reduces to the well-known result

$$
\begin{pmatrix} A & B \\ C & D \end{pmatrix} = \begin{pmatrix} 1 & l_e/n \\ 0 & 1 \end{pmatrix}
$$
(2.15)

where l_e is the thickness of the etalon.

In addition to telescopes, gratings, multiple-prism arrays and etalons, it is useful to consider the active dye region in matrix form. This is certainly the most uncertain aspect of the use of matrix notation in dye laser cavities. The difficulty originates in the lack of characterization of the active region as an optical component and in the many physical factors, which can seriously alter any such characterization. In general, the active region can be described as

$$
\begin{pmatrix} A & B \\ C & D \end{pmatrix} = \begin{pmatrix} \alpha & \beta \\ \chi & \delta \end{pmatrix} .
$$
(2.16)

Certainly, the simplest description of a dye cell is that in which the dye is considered to provide a passive refractive index (n_d) region for a given active medium length (t). In this case $\alpha = \delta = 1$, $\chi = 0$, and $\beta = t/n_d$. Thermal lensing is known to affect beam divergence characteristics in laser-pumped pulsed dye lasers [2.40, 41]. If the active region is considered to be a convex (positive) lens, then (2.16) becomes (2.2). For the case of a concave (negative) lens $\chi = 1/|f|$. Description of the active length by a generalized series of lenses forming a cylindrical region can be made using the results given by *Kogelnik* [2.33].

It should be noticed that thermal lensing occurs both in the steady state and transient regime. In the steady state thermal lensing can be the result of environmental factors [2.41] or due to heating of the dye by the pump laser, as in the case of high pulse repetition frequency (prf) operation. Thermal lensing in the transient regime [2.42, 43] often results from the use of large pump pulse energies.

The description of the active length in the case of flashlamp-pumped dye lasers is considerably more complex since the medium is made optically in-homogeneous by radial temperature gradients exacerbated by flow turbulence [2.44].

Additional matrix notation for optical systems include larger matrices of the 3×3, 4×4, and 6×6 format; see, for example, [2.35, 45]. *Siegman* [2.35] describes 3×3 matrices such as

$$\begin{pmatrix} A & B & E \\ C & D & F \\ 0 & 0 & 1 \end{pmatrix} \qquad (2.17)$$

where the $ABCD$ terms have their usual meaning and the F term can be related to dispersion [2.46]. A detailed description of intracavity dispersion is given in Sect. 2.2.2. In the 4×4 notation the matrix can take the form of [2.47]

$$\begin{pmatrix} A & B & 0 & E \\ C & D & 0 & F \\ G & H & 1 & I \\ 0 & 0 & 0 & 1 \end{pmatrix}. \qquad (2.18)$$

Again, in this notation the F term can be related to dispersion. An alternative way to express the 4×4 matrix is to use the notation discussed by *Siegman* [2.35] where the 4×4 matrix is composed of four 2×2 matrices with each of the usual $ABCD$ terms at the upper left corner of the corresponding 2×2 matrix. Thus the 4×4 matrix can be written as

$$\begin{pmatrix} A & B \\ C & D \end{pmatrix} \qquad (2.19)$$

in shorthand notation [2.35]. Here the $\boldsymbol{ABCD}$ are submatrices given by [2.47, 48]

$$A = \begin{pmatrix} A & 0 \\ G & 1 \end{pmatrix}$$

$$B = \begin{pmatrix} B & E/\lambda_0 \\ H & I/\lambda_0 \end{pmatrix}$$

$$C = \begin{pmatrix} C & 0 \\ 0 & 0 \end{pmatrix}$$

$$D = \begin{pmatrix} D & F/\lambda_0 \\ 0 & 1 \end{pmatrix} .$$

λ_0 is defined as in the Huygens integral [2.48].

As stated previously the $ABCD$ terms in this notation have their usual meaning and F is related to dispersion. For the mth prism the following relations may be useful to identify and provide a physical meaning to other matrix components

$$E_m = B_m k_{1, m} F_{\psi_{2, m} = 0} \tag{2.20}$$

$$G_m = F_m A_m / \lambda_0 \tag{2.21}$$

$$H_m = B_m F_{\psi_{1, m} = 0} (1/\lambda_0) \tag{2.22}$$

where $F_{\psi_{2, m} = 0}$ represents the dispersion component for orthogonal beam exit and $F_{\psi_{1, m} = 0}$ is the dispersion component for orthogonal beam incidence. It should also be noted that the G_m term includes the single-pass dispersion multiplied by the beam magnification factor A_m and thus may be associated with the return-pass dispersion.

For a multiple-prism array the $ABCD$ terms are as given in (2.7, 8), the F term is related to the single-pass version of (2.27) and the G term can be related to (2.27) multiplied by $(2\lambda_0)^{-1}$. Thus, the generalized G coefficient can be written in implicit form as

$$G_r = \sum_{m = 1}^{r} F_m \left[\prod_{j = 1}^{m} k_{1, j} \prod_{j = 1}^{m} k_{2, j} \right] 1/\lambda_0 .$$

The generalized E, H, and I terms are too extensive to be included in the text.

In addition to resonator stability analysis the ray matrices presented here can be utilized to describe the beam divergence characteristics of the cavities. Utilizing the approach introduced by *Littman* and *Metcalf* [2.9] and the usual propagation equation for a Gaussian beam in an $ABCD$ system [2.38], it can be shown that the convolution value for the beam divergence can be written as

$$\Delta\theta = \frac{w}{L_R} \left[\left(\frac{L_R}{B'} \right)^2 + \left(\frac{L_R}{B'} \right)^2 (A\alpha + B\chi)^2 + \left(\frac{A\beta + B\delta}{B'} \right)^2 \right]^{1/2} \tag{2.23}$$

where $L_R = \pi w^2 / \lambda$ is the Rayleigh length. If the active length is characterized as a Gaussian aperture in conjunction with a lens [2.35], so that $\alpha = 1$, $\beta = 0$, $\chi = \chi$, and $\delta = 1$, then (2.23) reduces to [2.37]

$$\Delta\theta = \frac{w}{L_R} \left[1 + \left(\frac{L_R}{B} \right)^2 + \left[\frac{L_R(A + B\chi)}{B} \right]^2 \right]^{1/2} . \tag{2.24}$$

If we consider only the lens component of the active region then $\chi = -1/f$ for a positive lens and $\chi = 1/|f|$ for a negative lens. In either case, for a long focal

Table 2.3. Beam divergence as a function of cavity length

Cavity length	Beam divergence	Ref.
$d = L_R$	$\Delta\theta \approx \lambda\sqrt{3}/\pi w$	[2.32,[a] 37]
$d \ll L_R$	$\Delta\theta \approx \lambda\sqrt{2}/\pi w (L_R/d)$	[2.9[a]]
$d \gg L_R$	$\Delta\theta \approx \lambda/\pi w$	[2.13[a]]

[a] For these references the results were derived considering free space propagation.

length χ approaches zero and (2.24) becomes [2.37]

$$\Delta\theta = \frac{w}{L_R}\left[1 + \left(\frac{L_R}{B}\right)^2 + \left(\frac{L_R A}{B}\right)^2\right]^{1/2}. \tag{2.25}$$

In the absence of a beam expander, or other intracavity components, one can write $A = 1$, and $B = d$, so that (2.25) reduces to the well-known results listed in Table 2.3.

Duarte [2.37] compares the value of the measured $\Delta\theta$ with that calculated using (2.25) for a grazing incidence and a HMPGI cavity. For the HMPGI cavity the measured value was 1.33 ± 0.16 mrad as compared to the calculated value of 2.09 mrad. The measured $\Delta\theta$ is close to the diffraction limited value of ~ 1.2 mrad [2.37].

We should also note that possible astigmatic aberrations in multiple-prism cavities may be corrected by making one of the prism exit surfaces convex cylindrical [2.28].

For a discussion of beam divergence as related to intracavity apertures and the uncertainty principle $\Delta p \, \Delta x \approx h$, the reader should refer to Chap. 6.

2.2.2 Intracavity Dispersion

The basics of intracavity dispersion in multiple-prism grating oscillators have been treated in detail in [2.2]. Here we use the main results to describe this subject in a succinct manner.

The round-trip (or double-pass) generalized dispersion for a multiple-prism grating assembly composed of r prisms of any geometry or material is given by *Duarte* and *Piper* [2.26]

$$\left(\frac{d\Theta}{d\lambda}\right)_C = M\left(\frac{\partial\Theta}{\partial\lambda}\right)_G + \left(\frac{d\Phi}{d\lambda}\right)_P \tag{2.26}$$

where the prismatic dispersion is given by *Duarte* [2.2, 49]

$$\left(\frac{d\Phi}{d\lambda}\right)_P = 2\left(\prod_{j=1}^{r} k_{1,j} \prod_{j=1}^{r} k_{2,j}\right) \sum_{m=1}^{r} (\pm 1)\frac{\tan\phi_{1,m}}{n_m}\frac{dn_m}{d\lambda}\left(\prod_{j=m}^{r} k_{1,j} \prod_{j=m}^{r} k_{2,j}\right)^{-1}$$

$$+ 2\sum_{m=1}^{r} (\pm 1)\frac{\tan\phi_{2,m}}{n_m}\frac{dn_m}{d\lambda}\prod_{j=1}^{m} k_{1,j} \prod_{j=1}^{m} k_{2,j}. \tag{2.27}$$

Here $k_{1,j} = \cos\psi_{1,j}/\cos\phi_{1,j}$, $k_{2,j} = \cos\phi_{2,j}/\cos\psi_{2,j}$, and

$$M = \prod_{j=1}^{r} k_{1,j} \prod_{j=1}^{r} k_{2,j} .$$

It should be noted that to obtain the single-pass multiple-prism dispersion (2.27) is multiplied by

$$\left(2\prod_{j=1}^{r} k_{1,j} \prod_{j=1}^{r} k_{2,j}\right)^{-1} \tag{2.28}$$

to provide $d\phi_{2,r}/d\lambda$ as given in [2.2, 49]. This latter expression can be related to the F term of the 4×4 matrix given in (2.18, 19). For orthogonal beam exit (2.27) reduces to the well-known result given in [2.50, 51].

In this notation the double-pass dispersive linewidth can be expressed as

$$\Delta\lambda \approx \Delta\theta\left[M\left(\frac{\partial\Theta}{\partial\lambda}\right)_G + \left(\frac{d\Phi}{d\lambda}\right)_P\right]^{-1} \tag{2.29}$$

wnere $(\partial\Theta/\partial\lambda)_G$ is the grating dispersion either in Littrow or grazing incidence configuration [2.2]. *Duarte* and *Piper* [2.52] provide an expression for the linewidth in the format of (2.29) that describes multipass dispersive effects.

For several reasons it is sometimes desirable to design multiple-prism beam expanders yielding zero dispersion at a given wavelength. This type of beam expander is referred to as achromatic or quasi-achromatic in the literature [2.50, 53, 54]. In general, one can design an infinite number of dispersionless multiple-prism expanders by making $(d\Phi/d\lambda)_P = 0$ in (2.27).

2.2.3 Intracavity Transmission Efficiency

For a prismatic array the cumulative reflection losses at the incidence face are given by [2.2, 25]

$$L_{1,m} = L_{2,m-1} + (1 - L_{2,m-1})R_{1,m} \tag{2.30}$$

and the losses at the exit face are given by [2.25]

$$L_{2,m} = L_{1,m} + (1 - L_{1,m})R_{2,m} \tag{2.31}$$

where the factors $R_{1,m}$ and $R_{2,m}$ are given by the usual Fresnel equations for s- and p-polarization [2.55].

The efficiency of gratings can be determined experimentally in the format used by *Duarte* [2.24] or theoretically using electromagnetic theory as suggested by *Maystre* [2.56].

2.2.4 Fabry-Pérot Etalons

In addition to gratings and prisms, intracavity Fabry-Pérot etalons are widely used to provide fine frequency tuning and further linewidth narrowing. For a comprehensive review of Fabry-Pérot etalons the reader should refer to *Born* and *Wolf* [2.55], *Steel* [2.57], or *Meaburn* [2.58]. In this section we provide a basic introduction to the notation and parameters of interest. An important parameter in Fabry-Pérot etalons or interferometers is the free spectral range FSR defined as

$$\text{FSR} = \frac{c}{2nl_e} \tag{2.32}$$

in frequency units (Hz). Here n is the refractive index, and l_e the distance between the reflective surfaces. The $\text{FSR} = \lambda^2/2nl_e$ in wavelength units (m) and is equal to $1/2nl_e$ in wave numbers (m^{-1}).

In addition to the FSR another parameter of importance is the finesse ($\mathscr{F}$) of the etalon. This parameter depends on the flatness of the surfaces (often in the $\lambda/50 - \lambda/100$ range or better), the size of the aperture, and the reflectivity of the surfaces. The effective finesse is given by [2.58]

$$\mathscr{F}^{-2} = \mathscr{F}_R^{-2} + \mathscr{F}_F^{-2} + \mathscr{F}_A^{-2} \tag{2.33}$$

where $\mathscr{F}_R$, $\mathscr{F}_F$, and $\mathscr{F}_A$ are the reflective, flatness and aperture finesses, respectively. The reflective finesse is given by [2.55]

$$\mathscr{F}_R = \frac{\pi\sqrt{R}}{1 - R} \tag{2.34}$$

and the minimum resolvable bandwidth can be written as

$$\Delta(\text{FSR}) = \frac{\text{FSR}}{\mathscr{F}} . \tag{2.35}$$

The tuning characteristics of the etalon can be described using [2.55]

$$m\lambda = 2nl_e \cos\psi_e \tag{2.36}$$

where l_e is the thickness and ψ_e is the corresponding angle of refraction.

Multiple-etalon systems are described in detail by *Maeda* et al. [2.7] and *Pacala* et al. [2.59]. These authors provide the FSR and finesse of each intracavity etalon in their multiple-etalon assembly. In general, these systems are designed so that the FSR of the successive etalon is compatible with the measured laser linewidth obtained with the previous etalon or etalons.

2.3 Architecture and Mechanics

In this section we examine some issues that are important to optimize the lasing characteristics of dispersive resonators. In this context, we consider the architectural and mechanical elements necessary to integrate into the design the qualities of compactness, ruggedness, high conversion efficiency, low amplified spontaneous emission (ASE), thermal and dye-flow stability.

2.3.1 Physical Requirements on Architecture

A key ingredient in a successful design is compactness. A short, compact cavity offers a wide intracavity mode separation (given by $c/2L$, where L is the cavity length) which in turn facilitates the task of achieving single-longitudinal-mode oscillation. Thus, it is important to minimize L in an efficient and convenient manner. Here we describe how to mount the optical components, so that in addition to a minimum use of intracavity space, the optical components themselves are utilized to their maximum.

Firstly, we consider a laser-excited MPL oscillator as depicted in Fig. 2.2 with the beam expander schematics given in Fig. 2.3c or d. The following features should be highlighted:

i)   The width of the dye cell is ~ 10 mm;
ii) The distance from the output coupler to the dye cell is < 5 mm;
iii) The distance from the dye cell to the first prism of the beam expander is < 5 mm;
iv) The distance between prisms is 1–2 mm;
v)   The distance from the last prism of the expander to the grating is < 5 mm. However, this parameter may depend on tuning requirements.

In the case of prototype and research dye lasers the cavities are displaced over optical tables that provide a rigid surface and hence excellent mechanical stability. A practical alternative to deploy the different optical components of the resonator is to use good quality commercial kinematic mounts with ~ 0.1 sec of arc resolution or better. These mounts are themselves positioned over magnetic bases. This choice of mechanical components provides a very important freedom of displacement in two dimensions.

Once the height of the plane of propagation of the oscillator has been determined, appropriate intermediate metal bases are designed and fabricated to couple the kinematic mounts to the magnetic bases.

Commercial kinematics mounts usually employed offer ~ 52 mm apertures and are not well suited for the successful integration of a compact cavity. Thus, compactness and full utilization of optical surfaces requires the special design of intermediate optical mounts for mirrors and gratings. For example, in Fig. 2.4 a particular intermediate optical mount for a 140 mm holographic grating is illustrated. Similarly, multiple-prism beam expanders may necessitate the use of specific designs for their platforms with shapes appropriate to minimize demands on intracavity space. A particular design that we have found quite useful is a small optical table in the form of an elongated oval. Usually the metal used has a low thermal expansion coefficient. The mechanical approach described here offers a significant freedom of displacement that makes the possible use of alternative oscillator configurations, such as MPL and HMPGI, relatively easy.

GRATING MOUNT

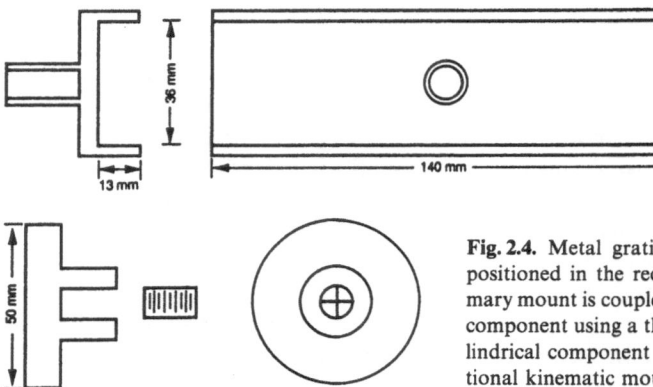

Fig. 2.4. Metal grating mount. The grating is positioned in the rectangular holder. This primary mount is coupled to the smaller cylindrical component using a threaded connector. The cylindrical component is positioned on the traditional kinematic mount

The design of the multiple-prism expander offers an additional opportunity for space reduction. In the case of narrow-linewidth laser-pumped dye lasers the ideal beam diameter is in the 200–400 μm range. Since several prisms are utilized in the expansion process the first and second prism can be relatively small and subsequent prisms can be made according to their position in the multiple-prism expander.

For flashlamp-excited oscillators the available beam diameter can be more than 5 mm. In this case, one can either use an aperture to reduce the beam dimensions to an appropriate size or a relatively small prism can be used as the first stage of magnification in the multiple-prism expander.

One further issue in the area of compactness is the balance between low intracavity losses and reduced intracavity length. As established in early experiments [2.20], the use of multiple-prism beam expansion, with prisms utilized at lower angles of incidence, reduces intracavity losses. The question here is whether to achieve $M = 100$ using eleven prisms at $\phi_{1,m}$ slightly higher than the Brewster angle, or using three or four prisms at higher angles of incidence. Fortunately, the high gain of dye lasers does not require the use of multiple-prism beam expanders with more than four stages of intracavity magnification.

Most of the details discussed here have been implicit features in experiments already reported; see, for example, [2.20, 22, 23, 37].

Constraints in commercial dye lasers and dye lasers designed for field deployment are much more severe than those applicable to research and prototype resonators. In the case of commercial dye lasers the availability of an optical table and two-dimensional displacement is often not an option. Here, optical mounts are usually attached to a linear structure using traditional mechanical methods. Certainly these designs are permanent and provide relatively fewer alternatives as compared to the research domain described previously. Constraints applicable to dye lasers designed for field deployment are even more severe since mechanical vibrations and environmental factors such as thermal variations become important variables. In this case rigid structures, using invar or superinvar, are installed on pneumatic supports. A ruggedized superinvar version of the narrow-linewidth flashlamp-pumped multiple-prism grating oscillator described in [2.25] has been found to deliver 2–3 mJ at $\Delta v \approx 300$ MHz and $\Delta\theta \leqslant 0.5$ mrad. Following vehicular displacement on a rugged terrain this ruggedized dispersive oscillator has demonstrated excellent performance as determined from emission parameters such as output laser energy, Δv, and $\Delta\theta$ [2.60].

2.3.2 Specifications for Optical Components

The proper design and deployment of optical components can play an important role in the improvement of conversion efficiency and reduction of ASE.

Firstly, as indicated in early publications (see, for example, [2.13, 20]) the dye cells should be of trapezoidal or parallelogrammatic geometry, as depicted in Figs. 2.1, 2, to reduce broadband parasitic internal reflections, and thus help to minimize the ASE. Notice that in addition to helping reduce ASE, this type of dye cell geometry enhances the propagation efficiency of the p-polarized intracavity energy flux. In addition, the use of broadband antireflection coatings may be used. These recommendations should be observed both in laser-pumped and flashlamp-excited devices.

In addition to the use of appropriate dye cell geometries careful attention to the transmission characteristics of multiple-prism beam expanders should be given. In Sect. 2.4 we discuss the design and transmission efficiency of multiple-prism beam expanders. In this context, broadband back reflections should be minimized by appropriate deployment of the prisms and by the use of antireflection coatings at the exit surfaces.

In the case of Galilean or Newtonian telescopic resonators, the telescopes can be deployed at a very slight angle to the optical axis to reduce the effect of back reflections. Antireflection coatings of the lens components are also employed. For the design of specific oscillators requiring a given beam magnification factor, it is important to select appropriate values for the focal length of the lenses to reduce the overall length of the telescope.

Gratings utilized in dye lasers are mainly *echelle* gratings and holographic gratings. An example of a particular echelle grating is that used by *Duarte* and *Piper* [2.20]. This grating has 632 lines/mm, blazed for a relatively high angle (54° 6'), and is used in Littrow configuration in the 5th or 6th order. Since this type of grating is 50 mm in diameter, its use in such high order reduces the requirements on intracavity beam expansion.

Holographic gratings of interest to dye lasers are produced in a rectangular format at 1800, 2400, 3000, 3600, and 4300 lines/mm. Usually their width is 50 mm but gratings up to 140 mm wide have been employed [2.23, 25]. These gratings are either used in Littrow configuration in the first order or in a grazing-incidence configuration.

The selection of a particular grating is mainly determined by the wavelength region of oscillation. For lasing in the near UV and blue portion of the spectrum, holographic gratings with 3600 and 4300 lines/mm are used. For operation in the visible to red portions of the spectrum, gratings with 2400 and 3000 lines/mm can be utilized. An important parameter for gratings is their efficiency at a particular spectral region. As stated earlier this can be determined experimentally [2.24] or theoretically [2.56]. Manufacturers often provide information on this parameter.

Materials utilized for dye cells, mirror substrates, prisms, and etalons are fused silica or high quality optical glass such as BK 7. An important requirement of these components is high surface quality with flatness figures approaching $\lambda/10$ or better. In the case of etalons where finesse depends on surface flatness, the flatness figure can approach $\lambda/100$ or better. Coatings for mirrors and etalons should be low-loss dielectric coatings.

2.3.3 Thermal and Dye Flow Control

Dispersive narrow-linewidth dye laser oscillators are high-performance tunable radiation sources which have an intrinsic sensitivity to certain physical phenomena. It is the task of the designer to incorporate configurational, mechanical and material features to neutralize the effects caused by variations in physical parameters such as environmental temperature and active medium optical homogeneity.

As demonstrated in a number of publications [2.40, 41] increase in the environmental temperature can induce increases in beam divergence, decrease in output power and shifting of the resonant laser frequency. These changes are clearly illustrated by *Duarte* [2.41] under severe and extreme conditions by increasing the environmental temperature from 20 to 35°C. However, the fact remains that for narrow-linewidth oscillation ($\Delta v < 1$ GHz) even a change in temperature of $\Delta T \sim 1$°C can induce serious frequency detuning. In order to minimize the effect of thermal variations several approaches should be taken:

i)   Utilize cavity structures of invar or super invar;
ii) Design zero-dispersion (or achromatic) multiple-prism beam expanders. See (2.27) and Sect. 2.4;
iii) If necessary, provide thermal control of the cavity temperature and maintain this temperature 1–2°C above room temperature;
iv) Circulate the dye in a system incorporating a heat exchanger;
v)   In the case of flashlamp-pumped dye lasers, control and maintain the ΔT between the cooling fluid and the dye solution to less than 0.1°C.

Normally only i) and ii) are necessary (and v) in the case of flashlamp-pumped dye lasers). The degree of implementation of these measures depends on various parameters such as prf, pulse energy of excitation laser, environmental temperature and even humidity. For instance, in hot humid climates (such as in the South Eastern United States or parts of Australia) condensation may occur on the dye cell if the dye is maintained at a temperature lower than the ambient temperature. This will cause laser oscillation to cease.

Effects on beam divergence due to thermal lensing can be described using (2.24) [2.2]. Consideration of this effect in the transient regime has been reported in [2.42, 43].

Dye flow turbulence can seriously affect the optical homogeneity of the active medium, and thus parameters such as beam uniformity and laser linewidth. *Duarte* and *Piper* [2.22] report that in order to achieve stable single-longitudinal-mode oscillation ($\Delta v \approx 650$ MHz) in their high-prf copper-laser-pumped HMPGI dye laser oscillator, dye flow had to be carefully controlled. When turbulence was eliminated and a dye laminar flow established then the HMPGI oscillator provided the time-integrated linewidth cited above.

On the other hand, thermal gradients in the active medium in conjunction with dye flow turbulence can induce dynamic linewidth instabilities as reported by *Duarte* and *Conrad* [2.23] and *Duarte* et al. [2.44]. In these experiments flashlamp-excited MPL and HMPGI oscillators yielding $\Delta v \leqslant 375$ MHz were utilized. As reported by these authors, thermal and dye-flow control can help in neutralizing these instabilities. Other authors [2.61, 62] have reported optical inhomogeneities in the active medium resulting from shock waves in long-pulse (~ 10 µs) flashlamp-pumped broadband dye lasers.

2.3.4 ASE Reduction

The level of amplified spontaneous emission (ASE) is an important parameter for many applications including lidar, spectroscopy and laser isotope separation (LIS). Here, we define the level of ASE using the spectral density concept described by *Duarte* [2.2]

$$\varrho_{ASE}/\varrho_l \approx \left[(\Delta\Lambda)^{-1} \int_{\Lambda_1}^{\Lambda_2} W(\Lambda)d\Lambda \right] \Big/ \left[(\Delta\lambda)^{-1} \int_{\lambda_1}^{\lambda_2} E(\lambda)d\lambda \right] \tag{2.37}$$

where $\Delta\Lambda$ is the full width of the broadband ASE emission and $\Delta\lambda$ is the laser linewidth at full width. For single-longitudinal-mode emission $E(\lambda)$ approaches the form of a Gaussian distribution. This definition of ASE takes into account the spectral brightness of the laser. For the ideal case of identical ASE and laser energy distributions, (2.37) reduces to the ratio of their respective maximum intensities $\{I_{ASE}/I_l\}$. Also, if we multiply (2.37) by $(\Delta\Lambda/\Delta\lambda)$ we obtain an expression for the ASE percentage.

The ASE percentage can be measured taking advantage of the highly divergent characteristics of ASE using geometrical methods [2.63]. The spectral density method requires the use of a spectrometer as described by *Duarte* and *Piper* [2.20] and other authors [2.64–67]. An additional method of determining ASE in a relatively narrow spectral region is the use of atomic or molecular absorption.

Table 2.4. ASE characteristics of multiple-prism grating oscillators[a]

Excitation source	E_0 [mJ]	Δv [MHz]	I_{ASE}/I_l	ϱ_{ASE}/ϱ_l	C[M]	Ref.
Coaxial	2.6–4.2	$\leqslant 360$	7×10^{-9}	1.2×10^{-9}	1.25×10^{-5}	[2.25]
Coaxial	~ 3	$\leqslant 360$[b]	6×10^{-10}	1.7×10^{-10}	1.25×10^{-5}	[2.25]
Linear	3.6	$\leqslant 138$	6×10^{-11}	2.9×10^{-11}	1×10^{-5}	[2.25]
CVL		60[c]	2×10^{-7}		6×10^{-4}	[2.21]
CVL		~ 400	5×10^{-7}		2×10^{-3}	[2.22]
Nitrogen laser		1610	7×10^{-6}		$\sim 10^{-2}$	[2.20]

[a] Adapted from [2.25].
[b] Incorporates intracavity polarizer.
[c] Incorporates intracavity etalon.

Reduction of ASE levels can be accomplished in a number or a combination of techniques:

i) Use closed cavity configurations as described by *Duarte* and *Piper* [2.20, 22];
ii) Reduce and modify surfaces giving origin to broadband back reflections. This is particularly important in dye cells and telescopes (see Sect. 2.3.2). The use of antireflection coatings is recommended;
iii) Deploy orthogonal multiple-prism beam expanders so that $\phi_{2,m} \approx 0$ rather than $\phi_{2,m} = 0$. This type of deployment can help to reduce ASE and has little effect on dispersive properties. The use of specifically designed non-orthogonal exit multiple-prism beam expanders is a better approach (see Sect. 2.4). Again the use of anti-reflection coatings is recommended here;
iv) For a given excitation laser and type of dye, select an appropriate dye concentration so that the value given by (2.37) is as little as possible. This is particularly important for the oscillator stage;
v) Use delayed excitation of amplifier stages as described by *McKee* et al. [2.63] and *Dupre* [2.32];
vi) Take advantage of the broadband and high divergence characteristics of ASE using appropriate spectral and spatial filters as described by *Wallenstein* and *Hänsch* [2.68].

In a recent set of experiments *Duarte* et al. [2.25] utilized MPL and HMPGI flashlamp-pumped dye lasers oscillators to demonstrate $\{\varrho_{ASE}/\varrho_l\} \approx 10^{-9}$ at linewidths in the $138\,\text{MHz} \leqslant \Delta v \leqslant 360\,\text{MHz}$ range. This was accomplished using fairly low dye concentrations ($\sim 10^{-5}$ M) and a Glan-Thompson polarizer at the output coupler end of the cavity. In Table 2.4 we provide ASE figures for a number of narrow-linewidth multiple-prism grating oscillators.

2.3.5 Wavelength Tuning

Wavelength tuning in pulsed dye lasers has been described in detail in previous publications [2.1, 3]. Most wavelength tuning techniques use the inherent angular wavelength-dependent characteristics of cavity components such as prisms, etalons and gratings to vary the resonant frequency.

Accurate angular displacement of tuning mirrors and/or gratings requires employment of high quality kinematic mounts with < 0.1 s of arc resolution. In this context, it should be mentioned that a displacement of the resonant frequency of $\delta v \sim 250\,\text{MHz}$, equivalent to the Δv of a flashlamp-pumped MPL oscillator, requires an angular rotation of $\delta\Theta \approx 10^{-6}$ rad at the grating [2.44]. In addition to angular resolution, desired features of kinematic mounts used in frequency tuning include mechanical precision and tight angular tolerances. In other words, the setting in the vernier micrometer should correspond to a given angular position regardless of the direction of the motion of the vernier micrometer.

Motion of kinematic mounts can be accomplished by careful hand rotation of vernier micrometer knobs or by displacement of mechanical or piezoelectric pushers driven by electronic controls. In this case, we should mention that relatively fast piezoelectric drivers providing a resolution in the sub-micrometer range, are commercially available.

In addition to angular displacement, intracavity Fabry-Pérot etalons can be tuned utilizing pressure tuning techniques as described by *Wallenstein* and *Hänsch* [2.69]. For a description of synchronous tuning in single-mode grazing-incidence dye lasers the reader should refer to [2.11].

2.4 Multiple-Prism Beam Expander Configurations: Examples

In this section we utilize the generalized dispersion equations presented in Sect. 2.2.2 to discuss the design of various multiple-prism expanders. Firstly, we should emphasize that the generalized dispersion equations given in [2.2, 26, 49, 50] can be used to design an infinite variety of multiple-prism expanders of any material and geometry. As an introductory step we consider multiple-prism expanders designed for orthogonal beam exit. In this case $\phi_{2,1} = \phi_{2,2} = \ldots = \phi_{2,m} = 0$ so that $k_{2,1} = k_{2,2} = \ldots = k_{2,m} = 1$, and (2.27) reduces to the well-known result [2.50, 51]

$$\left(\frac{d\Phi}{d\lambda}\right)_{\text{P}} = 2 \sum_{m=1}^{r} (\pm 1)\left(\prod_{j=1}^{m} k_{1,j}\right) \tan\psi_{1,m}\frac{dn_m}{d\lambda} \tag{2.38}$$

and

$$M = \prod_{m=1}^{r} k_{1,m} .$$

Furthermore, for identical prisms, orthogonal beam exit, and incidence at the Brewster angle so that $k_{1,1} = k_{1,2} = \ldots = k_{1,m} = \ldots = n$, (2.27) takes the elegant form given in [2.2]

$$\left(\frac{d\Phi}{d\lambda}\right)_P = 2 \sum_{m=1}^{r} (\pm 1)(n)^{m-1}\frac{dn}{d\lambda}$$

(2.39)

and

$$M = n^r .$$

(2.40)

Again, to obtain the single-pass multiple-prism dispersion (2.39) is multiplied by $(2n^r)^{-1}$ and the result is that given in [2.2].

Initially, we illustrate some basic characteristics of multiple-prism expanders utilizing simple additive configurations. In Table 2.5 we provide, for $M = 30$, the calculated single-pass transmission factor (T) and the dispersion for expanders with two, three, and four prisms. The transmission factor is found using (2.30, 31) and the material considered in all these examples is fused silica at 580 nm ($n = 1.458734$ and $dn/d\lambda = -3.646086 \times 10^4$ m^{-1}). The basic information provided in Table 2.5 is that, for a fixed M, the absolute value of dispersion increases and the transmission increases as the number of prisms in the additive configuration increases. Since $|(d\Phi/d\lambda)_P|$ is only a small percentage of the overall dispersion dominated by the grating contribution ($M(\partial\Theta/\partial\lambda)_G$) [2.50], the important factor to consider is T. The problem to the designer is to provide the shortest possible cavity using a minimum number of prisms without introducing significant losses. The answer to this question depends on additional information on gain of the active medium and excitation parameters. The reader should

Table 2.5. Dispersion and transmission characteristics for multiple-prism beam expanders in additive configurations at $M = 30$

Number of prisms	$\left(\dfrac{d\Phi}{d\lambda}\right)_P$ [m^{-1}]	T
2	-2.3950×10^6	0.44
3	-2.7799×10^6	0.65
4	-3.1434×10^6	0.79

Table 2.6. Dispersive and transmission characteristics of additive $(+, +, +, +)$ multiple-prism expanders

$\phi_{1,1} = \phi_{1,3} = \phi_{1,3} = \phi_{1,4}$	M	$\left(\dfrac{d\Phi}{d\lambda}\right)_P$ [m^{-1}]	T
71.0128	30	-3.1434×10^6	0.79
76.3590	100	-9.4312×10^6	0.55
77.7518	150	-1.3726×10^7	0.47

notice that the magnification factor selected here is suitable for HMPGI oscillators; see, for example, [2.22].

In Table 2.6 the dispersive and transmission characteristics of four-prism expanders of the additive configuration $(+,+,+,+)$ are compared for $M = 30$, 100, and 150. For these expanders $\phi_{1,1} = \phi_{1,2} = \phi_{1,3} = \phi_{1,4}$ so that $k_{1,1} = k_{1,2} = k_{1,3} = k_{1,4}$. As the beam magnification factor increases $|(d\Phi/d\lambda)_P|$ increases and T decreases. The large M factors are utilized in MPL oscillators.

Table 2.7 lists dispersion and transmission values for eight different configurations of four-prism expanders. The dispersive and transmission figures of all these configurations (shown in Fig. 2.5) are determined by the basic zero-dispersion design of the $(+,+,+,-)$ configuration. Here, the first three prisms are identical, for example, at $M = 100.19$, $\phi_{1,1} = \phi_{1,2} = \phi_{1,3}$ $= 79.227359°$ $(k_{1,1} = k_{1,2} = k_{1,3} = 3.955)$ and the incidence angle of the compensating prism is $\phi_{1,4} = 60.253257°$ $(k_{1,4} = 1.619582)$. The reader should note that under these circumstances compensating configurations of the type $(+,-,+,-)$ and $(+,-,-,+)$ reduce the dispersion but do not eliminate it. If we specifically design a dispersionless multiple-prism beam expander of the $(+,-,+,-)$ or $(+,-,-,+)$ configuration for $M \approx 100$ with two identical pairs providing $k_{1,1} k_{1,2} = 10.062057$ $(\phi_{1,1} = \phi_{1,3} = 84.128586$ $(k_{1,1} = k_{1,3} = 7.149999)$ and $\phi_{1,2} = \phi_{1,4} = 53.673873$ $(k_{1,2} = k_{1,4} = 1.407281))$ then the overall transmission becomes $T = 0.31$. Thus, for zero dispersion configurations $T_{(+,-,+,-)} < T_{(+,+,+,-)}$. Information provided in Tables (2.5, 6) in conjunction with the transmission factors given above indicate that the highest transmissions are obtained when all or most prisms are identical and utilized at the same angle of incidence. This feature was implicit in earlier publications by *Duarte* and *Piper* [2.26] and was well illustrated by *Trebino* [2.54]. In general, optimum transmissions are obtained in configurations where all prisms are oriented at the same angle of incidence and $k_{1,1} = k_{1,2} = \ldots = k_{1,m}$

Table 2.7. Dispersive and transmission characteristics of four-prism laser beam expanders

Configuration	$\left(\dfrac{d\Phi}{d\lambda}\right)_P$ [m^{-1}]		
	$M = 30.95$ ($T = 0.76$)	$M = 100.19$ ($T = 0.48$)	$M = 151.74$ ($T = 0.39$)
I $(+,+,+,+)$	-3.4685×10^6	-1.0823×10^7	-1.6200×10^7
II $(+,+,+,-)$	0.0000	0.0000	0.0000
III $(+,+,-,+)$	-1.1995×10^6	-2.6037×10^6	-3.4109×10^6
IV $(+,+,-,-)$	2.2689×10^6	8.2195×10^6	1.2789×10^7
V $(+,-,+,+)$	-2.6008×10^6	-8.7449×10^6	-1.3402×10^7
VI $(+,-,+,-)$	8.6769×10^5	2.0782×10^6	2.7985×10^6
VII $(+,-,-,+)$	-3.3181×10^5	-5.2547×10^5	-6.1237×10^5
VIII $(+,-,-,-)$	3.1367×10^6	1.0297×10^7	1.5588×10^7

Here $\phi_{1,1} = \phi_{1,2} = \phi_{1,3}$ for all configurations. Dispersion compensation is provided by the fourth prism in the $(+,+,+,-)$ configuration.

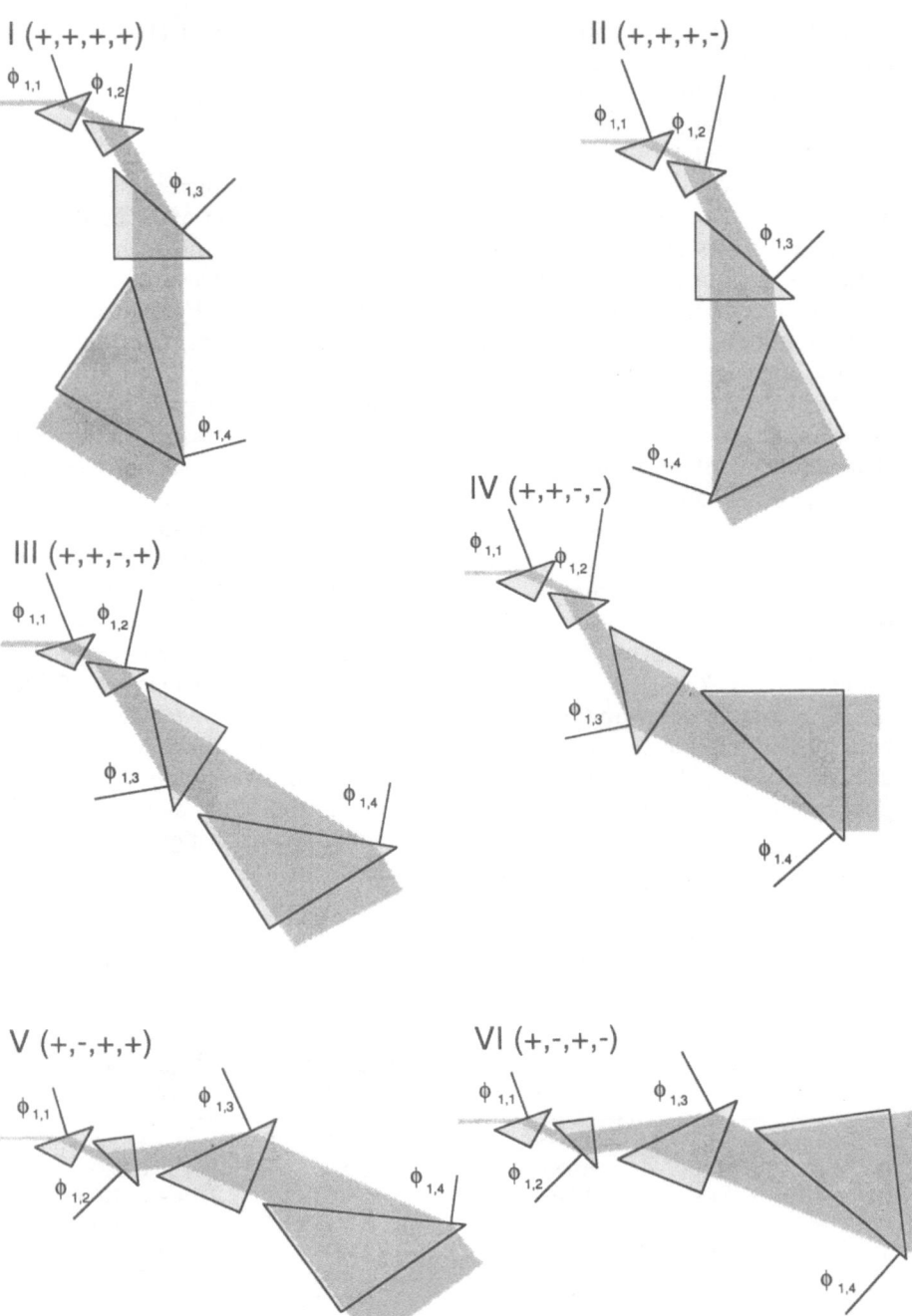

Fig. 2.5. For caption see opposite page

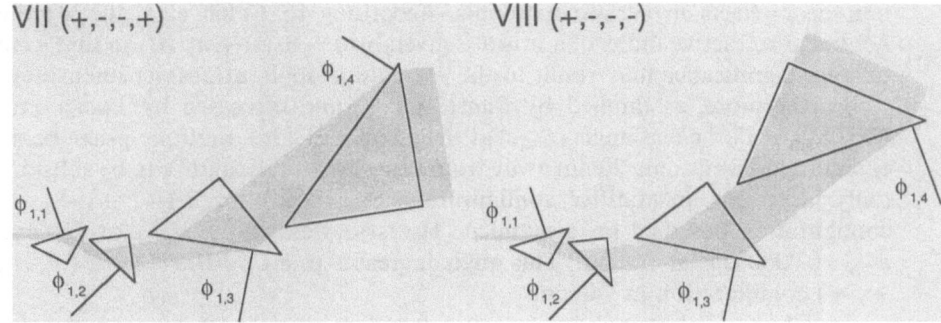

VII (+,-,-,+)

VIII (+,-,-,-)

Fig. 2.5. Orthogonal beam exit multiple-prism configurations for $r = 4$

so that

$$M = (k_{1,m})^r \ . \tag{2.41}$$

Unfortunately, under these circumstances dispersion can be reduced but not eliminated using compensating configurations.

As stated previously for a given beam magnification factor

$$M\left(\frac{\partial \Theta}{\partial \lambda}\right)_G \gg \left|\left(\frac{d\Phi}{d\lambda}\right)_P\right| \tag{2.42}$$

so that the prismatic dispersion has little influence on the linewidth value of the multiple-prism grating oscillator [2.50]. In this context the importance of zero dispersion (or achromatic beam expanders) is to reduce deviations of the expanded beam due to thermal variations or oscillation away from the wavelength of design.

An example of the usefulness of dispersionless multiple-prism beam expanders can be made considering single-pass angular deviations due to thermal gradients. The exit angle at the mth prism varies, as a function of temperature, according to [2.2]

$$\frac{d\phi_{2,m}}{dT} = \frac{d\phi_{2,m}}{d\lambda}\left(\frac{dn_m}{d\lambda}\right)^{-1}\frac{dn_m}{dT} \ . \tag{2.43}$$

It is easy to see that for a given non-zero value of $d\phi_{2,m}/d\lambda$ thermal variations, due to environmental factors or resulting from a high intracavity energy flux, can result in beam deviations that affect the resonant frequency. Thus, it is clear from this perspective that $d\phi_{2,m}/d\lambda \approx 0$ is a desirable design feature. For further details the reader should refer to [2.2, 50, 54]. An additional factor that may cause variations in the refractive index of the prisms is intensity-dependent nonlinear effects in certain materials. According to *Golub* et al. [2.70] the nonlinear refractive index of a prism is given by $n = n_0(v) + n_2 I(t)$ so that a Δn of some significance may result in the presence of high intracavity intensities.

Furthermore, as implied by *Duarte* [2.49] and discussed by *Duarte* and *Conrad* [2.71] the exit angle ($\phi_{2,r}$) at the last prism, for a multiple-prism beam expander utilized some 20 nm away from its design wavelength, can be substantially larger for an additive configuration. Specifically for a $(+, +, +, -)$ configuration designed for alexandrite lasers to yield $d\phi_{2,r}/d\lambda = 0$ at 760 nm, $\phi_{2,r} = 0.0000054°$ at 740 nm. This angle increases to $\phi_{2,r} = 0.034°$ for a $(+, +, +, +)$ configuration at 740 nm.

In all the orthogonal beam exit expanders used in these examples we assumed that the exit face was antireflection coated so that the reflection coefficient is reduced to 0.3%. This is quite important to reduce broadband back reflections which can contribute to increasing the ASE level. As pointed out by *Duarte* [2.49], in practice only near-orthogonality is achieved ($\phi_{2,m} \approx 0$) so that the effect of these broadband reflections is reduced. A non-orthogonal multiple-prism beam expander in the $(+, +, +, -)$ compensating configuration specifically designed to minimize the effect of back reflections is described in Table 2.8. Several factors are worth mentioning here. We again utilized three identical prisms, giving $\phi_{1,1} = \phi_{1,2} = \phi_{1,3}$ and $\phi_{2,1} = \phi_{2,2} = \phi_{2,3}$ ($= 3.5177°$) in an additive configuration with a fourth prism in a compensating configuration (here $\phi_{2,4} = 2.8567°$). The transmission factor $T = 0.47$ is quite close to the transmission factor of the equivalent orthogonal multiple-prism expander (see Table 2.7). Also, if the last prism is rotated so that a $(+, +, +, +)$ is formed then $(d\Phi/d\lambda)_P = -1.166475 \times 10^7 \, \text{m}^{-1}$ which is similar to the $-1.0823 \times 10^7 \, \text{m}^{-1}$ value given for the orthogonal beam exit $(+, +, +, +)$ expander given in Table 2.7.

Table 2.8. Nonorthogonal exit multiple-prism achromatic beam expander in the configuration $(+, +, +, -)$ for $M = 103.1978$ and $T = 0.47$

$\phi_{1,1} = \phi_{1,2} = \phi_{1,3} = 79.3537°$

$\phi_{2,1} = \phi_{2,2} = \phi_{2,3} = 3.5177°$

$k_{1,1} = k_{1,2} = k_{1,3} = 4.0000$

$k_{2,1} = k_{2,2} = k_{2,3} = 0.9990$

$\phi_{1,4} = 60.2236°$

$\phi_{2,4} = 2.8567°$

$k_{1,4} = 1.6184$

$k_{2,4} = 0.9993$

The apex angle is given by $\alpha_m = \psi_{1,m} + \psi_{2,m}$.

2.5 Single-Longitudinal-Mode HMPGI Dye Laser Oscillator: Example

In this section we assume that we are given the task of designing a high performance narrow-linewidth oscillator. We apply some of the theory and information provided in previous sections to illustrate the design of a high prf (~ 10 kHz) single-longitudinal-mode oscillator for the orange-red region of the spectrum. A series of steps in the design process may be as follows:

i)   Determine the main wavelength region of operation. This indicates the type of dye to utilize;

ii)   Select the most efficient type of pump source to excite the dye in question. As discussed by *Duarte* [2.3], ideally the excitation source should have a wavelength of emission compatible with the $S_0 \rightarrow S_1$ absorption transition of the dye. Another very important parameter is prf;

iii) Design an appropriate dye cell;

iv)   Determination of the beam waist parameters: this determines the Rayleigh length of the cavity and its beam divergence characteristics. This also relates to the optimum cavity length;

v)   Decide on the type of oscillator configuration. If a resonator incorporating a beam expander is selected one should decide on the type of beam expander: telescope or multiple-prism expander. Grating type, configuration and dimensions should also be determined;

vi)   Design appropriate beam expander;

vii) Calculate the cavity length;

viii) Estimate return-pass dispersive linewidth;

ix)   If additional intracavity dispersion is needed, go back to v);

x)   The decision made in v) indicates the type of polarization characteristics of the resonator. For instance, we know that MPL and HMPGI oscillators emit almost 100% *p*-polarized light [2.22, 72]. If this is the case one should provide an appropriate polarization matching mechanism between the pump and the dye laser. This increases efficiency;

xi)   Minimize the value of ϱ_{ASE}/ϱ_l.

For our purpose we choose $\lambda \sim 580$ nm as the design wavelength. This requirement, in addition to the prf characteristics, limits the choice of excitation sources to either a recombination laser, such as Sr^+, or a CVL. Since efficiency for a red dye is better for excitation in the green, we select the CVL as the pump source. Having made this decision we next decide on the dye cell configuration. Using the information given in Sect. 2.3 we decide on a trapezoidal cell about 10 mm wide and $\leqslant 1$ mm thick. This narrow passage provides a flow cross-sectional area of $\leqslant 10$ mm^2 which is appropriate for a dye flow velocity of a few ms^{-1} utilizing a high-pressure dye pump, as discussed in [2.22]. Testing and measurements of flow characteristics to eliminate turbulence may be done at this stage.

The pump beam can now be focused at the cell to a thin and elongated shape ~ 10 mm wide with a cross section having a diameter in the $0.1-0.3$ mm range.

For $w \sim 0.15$ mm the Rayleigh length becomes $L_R \sim 12.2$ cm and the expected diffraction limited divergence $\Delta\theta \sim 1.23$ m rad.

Since single-longitudinal-mode oscillation is desired we must opt for a high dispersion compact cavity which limits our options to an oscillator of the MPL or HMPGI class. Now, we know that Littrow configurations are more efficient than near-grazing incidence configurations, but since CVL excitation of red dyes is known to be very efficient [2.42, 73] we can afford the use of a HMPGI configuration.

The grating we choose is a 3000 lines/mm holographic grating at an angle of incidence $\sim 80°$. This implies that the product $2wM \sim 9$ mm which can be accommodated using one of the multiple-prism designs providing $M \sim 30$. For this task we choose a four-prism expander in the $(+, +, +, -)$ configuration.

Notice that the largest prism needs an hypotenuse of about 15 mm only. For this expander $\phi_{1,1} = \phi_{1,2} = \phi_{1,3} = 73.2317°$ $(k_{1,1} = k_{1,2} = k_{1,3} = 2.615)$ and $\phi_{1,4} = 62.7312$ $(k_{1,4} = 1.7306)$ and its dimensions demand the use of no more than 5 cm of intracavity space. Thus the overall cavity length can be less than 12 cm. In order to match the Rayleigh length we choose a cavity length of ~ 12 cm and the cavity mode spacing becomes ~ 1.25 GHz.

The dispersion provided by the grating is calculated to be 3.4552×10^7 m^{-1}. Thus, using (2.29) the return-pass linewidth becomes $\Delta v \sim 1.83$ GHz. As discussed by *Bernhardt* and *Rasmussen* [2.21] and *Duarte* and *Piper* [2.52] only a couple of intracavity passes are necessary to restrict lasing to single-longitudinal-mode oscillation.

MPL and HMPGI oscillators are known to yield $\sim 100\%$ p-polarization [2.22, 72] thus it is advisable to match the polarization of the CVL to the preferred polarization of the cavity. This can be done by either rotating the orientation of the Brewster windows of the CVL, employing a colinear prismatic polarization rotator as described by *Duarte* [2.74] (see Fig. 2.6), or using multiple mirror reflections [2.75].

At this stage all the considerations for cavity rigidity and thermal stability should be made. Also, once the cavity is assembled and tested, a careful measurement of ϱ_{ASE}/ϱ_l should be made to achieve the lowest ASE figure. Here it should be indicated that the use of low dye concentrations provides increased mode discrimination [2.21] in addition to yielding low ASE values.

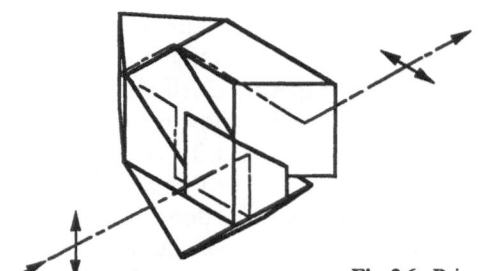

Fig. 2.6. Prismatic polarization rotator

2.6 Generation of Extremely Elongated Gaussian Beams
for Interferometry

In this section we briefly discuss the use of ray transfer matrices to characterize an optical system that produces extremely elongated Gaussian beams [2.76, 77]. For our purpose an extremely elongated Gaussian is defined as a beam whose ratio of maximum height to maximum width is 1:1000 or one part in a few thousand. Here we describe an optical system that produces beams with such characteristics and its application to interferometry.

The optical configuration is outlined in Fig. 2.7 and consists of a laser source providing a TM_{00} single-longitudinal-mode Gaussian beam, a Galilean telescope, a convex lens and a multiple-prism beam expander. Depending on the type of application, the laser can either be a single-transition source or a tunable laser such as a dye laser.

At the focal plane this optical system delivers an extremely elongated Gaussian beam which can be used to illuminate a variety of surfaces. A photodiode detector array located behind the experimental surface is then used to detect and record the transmission intensity. The instrument has two modes of operation. If the detector is placed immediately behind the experimental surface then the system is utilized as a classical microdensitometer. If, on the other hand, the detector is displaced a few centimeters from the experimental surface, interference occurs and the resulting interferometric signal is detected by the photodiode array. Typical detectors can incorporate 1024 or 2048 diodes each 25 µm wide.

The beam propagation characteristics of the optical system considered here can be easily described using (2.1–3, 7, 27, 30, 31). In Fig. 2.8 we show the intensity beam profile of an elongated Gaussian ($\sim 1:1000$ ratio) for a $5 \times$ multiple-prism magnification.

If the elongated Gaussian beam is utilized to illuminate a transmission grating, then the probability amplitude for a photon to go from the exit surface

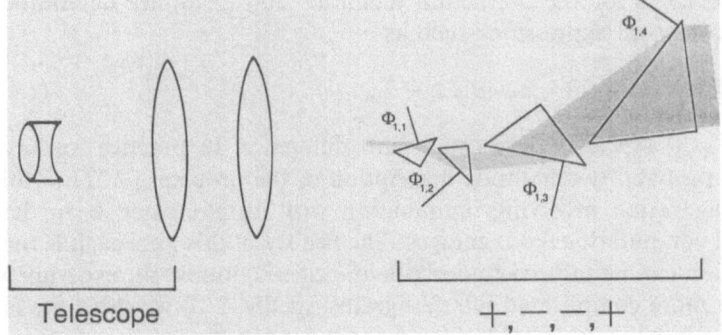

Fig. 2.7. Schematic of transmission telescope and multiple-prism system utilized to provide elongated Gaussian beam

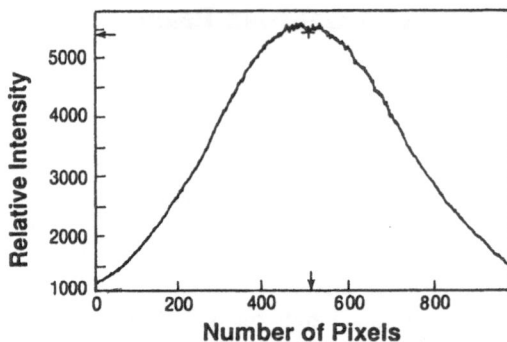

Fig. 2.8. Intensity profile of extremely elongated Gaussian beam 25 mm wide and 25 μm high (at its center)

of the multiple-prism expander (s) to the photodiode array (x), via an array of N slits (j), can be written as [2.78]

$$\langle x|s\rangle = \sum_{j=1}^{N} \langle x|j\rangle \langle j|s\rangle \ . \tag{2.44}$$

Following Feynman [2.79] we set $\langle j|s\rangle = \Psi(r_{s,j})\,e^{-i\theta_j}$ and $\langle x|j\rangle = \Psi(r_{j,x})$ $e^{-i\phi_j}$. Here quantities $\Psi(r_{s,j})$ and $\Psi(r_{j,x})$ are assumed to take the form of an appropriate diffraction wave function. Now if we write $\Psi(r_j) = \Psi(r_{s,j})\,\Psi(r_{j,x})$ and $\Omega_j = (\phi_j + \theta_j)$, then

$$\langle x|s\rangle = \sum_{j=1}^{N} \Psi(r_j)e^{-i\Omega_j} \ . \tag{2.45}$$

Hence the generalized probability expression for the one-dimensional N-slit problem is given by [2.77]

$$|\langle x|s\rangle|^2 = \sum_{j=1}^{N} \Psi(r_j)^2 + 2\sum_{j=1}^{N} \Psi(r_j)\left[\sum_{m=j+1}^{N} \Psi(r_m)\cos(\Omega_m - \Omega_j)\right] . \tag{2.46}$$

Angular relationships for the diffraction functions and (2.46) are determined using exact geometrical expressions such as

$$\cos(\phi_m - \phi_j) = \cos(\mathbf{k}\cdot\mathbf{r}) = \cos(k|L_m - L_{m-1}|) \tag{2.47}$$

where $|L_m - L_{m-1}|$ represents the exact path difference. In practice we have found that the probability amplitude description of the problem [2.77] in this particular configuration involving illumination with an elongated beam has conceptual and computational advantages. The beauty of this approach is that a simple equation can be utilized to describe the classic double slit experiment (see Fig. 2.9) or more complicated interferograms. In Fig. 2.10 we show the far field intensity pattern due to 100 slits (30 μm wide). Under these circumstances it should be noticed that in addition to interference $|\langle x|s\rangle|^2$ also provides information on the diffraction orders of the transmission grating.

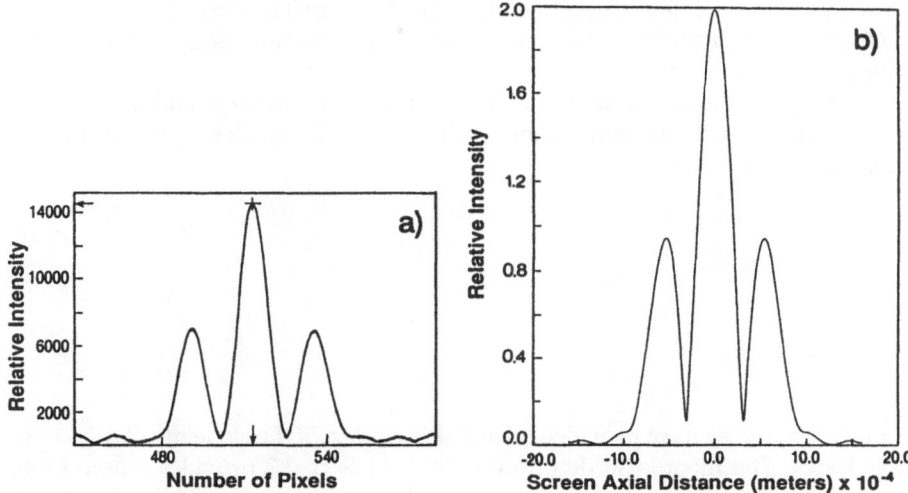

Fig. 2.9a, b. Classical double slit interference for slits 50 μm wide separated by 50 μm. (a) Measured intensity profile and (b) calculated interferometric intensity using (2.46) with $N = 2$. The distance between the slits (j) and the photodiode detector array (x) is 10 cm

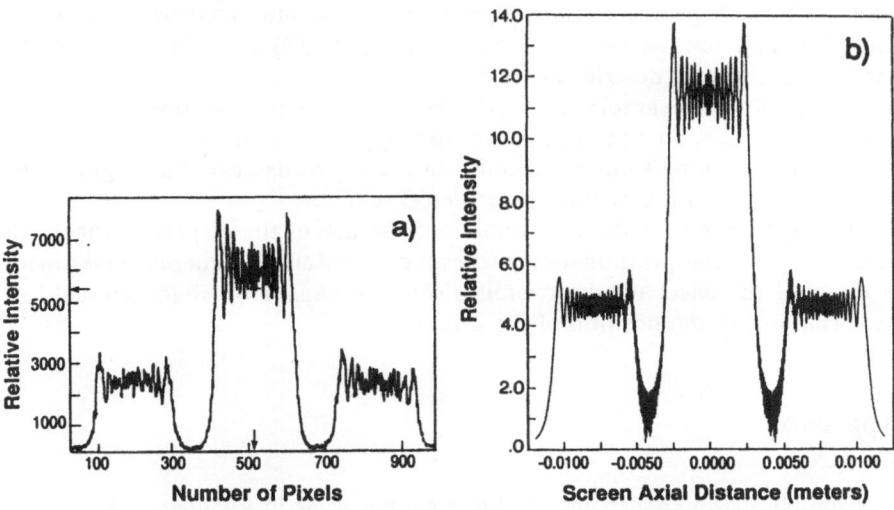

Fig. 2.10a, b. Multiple-slit interference for $N = 100$ using slits 30 μm wide separated by 30 μm. (a) Measured signal and (b) calculated intensity profile. The distance between the slits (j) and the photodiode detector array (x) is 75 cm

In practice, the device described here has been used to characterize a variety of transmission gratings, defects in clear film substrates and defects in thin metal films.

Finally, it should be indicated that using differentiation and subtraction, (2.47) can be used in conjunction with $\Delta k \, \Delta \lambda \approx 2\pi$ to derive the dispersive linewidth expression

$$\Delta \lambda \approx \Delta \theta \left(\frac{\partial \Theta}{\partial \lambda} \right)^{-1} . \tag{2.48}$$

2.7 Summary

In this chapter we have provided a basic description of the elements of dispersive dye lasers. The theoretical description given in Sect. 2.2 provides a useful link between the matrix notation and the dispersive theory, developed independently [2.2]. In addition, it should be highlighted that these dispersive and beam propagation equations can be applied to other gas and solid-state high-power laser oscillators.

Section 2.3 emphasized the technological nature of the design and construction of dispersive dye lasers. The interdependence with concepts described in Sect. 2.2 is also evident. The design examples of multiple-prism expanders (Sect. 2.4) and narrow-linewidth oscillators (Sect. 2.5) extensively utilize the theoretical elements described earlier.

An application example of both dye lasers and the propagation equations is given in Sect. 2.6. In this section we also apply the quantum description of interference resulting from a transmission grating to describe the origin of the single-pass linewidth equation $\Delta \lambda \approx \Delta \theta \, (\partial \Theta / \partial \lambda)^{-1}$.

The appendix provides an example of the use of the dispersive theory in prismatic pulse compression for femtosecond dye lasers. A numerical assessment is given of the effect on the overall dispersion caused by slight geometrical perturbations in the position of the prisms.

Appendix

A. Multiple-Prism Dispersion and Pulse Compression in Femtosecond Dye Lasers: Numerical Results

For completeness we quote the generalized single-pass dispersion given by *Duarte* [2.2, 80, 81]

$$\frac{d\phi_{2,m}}{dn} = \frac{\tan \phi_{2,m}}{n_m} \pm (k_{1,m} k_{2,m})^{-1} \left(\frac{\tan \phi_{1,m}}{n_m} \pm \frac{d\phi_{2,(m-1)}}{dn} \right) . \tag{2.49}$$

In most cases of interest the first sign is positive (see [2.2]) and the sign in the parenthesis becomes $(+)$ for additive configurations and $(-)$ for compensating configurations [2.80, 81]. For prismatic arrays, as utilized in pulse compression schemes, the apex angle $\alpha_m > \psi_{1,m}$ so that $d\psi_{1,m}/dn = -d\psi_{2,m}/dn$ and the second derivative given by [2.2, 80] can be reduced to

$$
\begin{aligned}
\frac{d^2 \phi_{2,m}}{dn^2} =\ & \tan \phi_{2,m} \left(\frac{d\phi_{2,m}}{dn} \right)^2 + (k_{2,m})^{-1} \tan \psi_{1,m} \tan \phi_{1,m} \frac{d\phi_{1,m}}{dn} \\
& \pm (k_{1,m} k_{2,m})^{-1} \frac{d^2 \phi_{2,(m-1)}}{dn^2} + \left(\frac{\tan \phi_{1,m}}{n_m} \pm \frac{d\phi_{2,(m-1)}}{dn} \right) \\
& \times \left\{ (n_m k_{1,m} k_{2,m})^{-1} (\tan \psi_{1,m} + \tan \psi_{2,m}) \right. \\
& \times \left. \left[(k_{1,m})^{-1} \frac{d\phi_{1,m}}{dn} - \tan \psi_{1,m} \right] - (k_{2,m})^{-1} n_m \tan \psi_{1,m} \frac{d\phi_{1,m}}{dn} \right\} .
\end{aligned}
\tag{2.50}
$$

This format of the derivative of $d\phi_{2,m}/dn$ has been found to be well suited for numerical calculations. Here the rates of change of the incidence and exit angles are related by

$$
\frac{d\phi_{1,m}}{dn} = \pm \frac{d\phi_{2,(m-1)}}{dn} ,
\tag{2.51}
$$

and this sign is positive for a compensating arrangement and negative for an additive configuration [2.80]. For this class of configuration the exit and incidence angle from prism to prism can be related by

$$
\phi_{1,m} = \phi_{2,(m-1)} \pm \Delta\zeta
\tag{2.52}
$$

for a compensating configuration [2.81] and

$$
\phi_{1,m} = \xi - \phi_{2,(m-1)} \pm \Delta\zeta
\tag{2.53}
$$

for an additive configuration. Here, ξ is the angle made by the intersection of the exit surface of the $(m-1)$th prism and the incidence surface of the mth prism. For the configuration discussed in [2.81] we have $\xi = 2\phi_w$, where ϕ_w is the corresponding Brewster angle of incidence. In (2.52, 53) $\Delta\zeta$ represents a small deviation of the prism from the position of a perfectly balanced configuration. For the simple case of a perfectly compensating configuration composed of two prisms pairs [2.82], at minimum deviations, and at the Brewster angle of incidence it can be shown that [2.81]

$$
\frac{d\phi_{2,1}}{dn} = \frac{d\phi_{2,3}}{dn} = 2
\tag{2.54a}
$$

$$\frac{d\phi_{2,2}}{dn} = \frac{d\phi_{2,4}}{dn} = 0 \tag{2.54b}$$

$$\frac{d^2\phi_{2,1}}{dn^2} = \frac{d^2\phi_{2,3}}{dn^2} = 4n - \frac{2}{n^3} \tag{2.54c}$$

$$\frac{d^2\phi_{2,2}}{dn^2} = \frac{d^2\phi_{2,4}}{dn^2} = 0 \tag{2.54d}$$

and the results for $m = 1$ and 2 are as given by [2.82]. It should be indicated that even slight deviations from a perfectly compensating configuration introduce a serious departure from the results given in (2.54). For example, if the second and fourth prism deviate by $\Delta\zeta = 1°$ from the compensating configuration we obtain (using numerical methods) [2.81]

$$\frac{d\phi_{2,1}}{dn} \approx 2.0000 \tag{2.55a}$$

$$\frac{d\phi_{2,2}}{dn} \approx -0.0396 \tag{2.55b}$$

$$\frac{d\phi_{2,3}}{dn} \approx 1.9611 \tag{2.55c}$$

$$\frac{d\phi_{2,4}}{dn} \approx 0.0777 \tag{2.55d}$$

$$\frac{d^2\phi_{2,1}}{dn^2} \approx 5.1813 \tag{2.55e}$$

$$\frac{d^2\phi_{2,2}}{dn^2} \approx -0.2831 \tag{2.55f}$$

$$\frac{d^2\phi_{2,3}}{dn^2} \approx 4.7308 \tag{2.55g}$$

$$\frac{d^2\phi_{2,4}}{dn^2} \approx 0.5408 \tag{2.55h}$$

for incidence at the Brewster angle (at $m = 1$) on quartz prisms, at $\lambda \sim 620$ nm. For further details of design and use of multiple-prism pulse compression the reader should refer to *Kafka* and *Baer* [2.83], *Duarte* [2.84] and *Diels* [2.85].

Information on prism parameters such as $n(\lambda)$ can be obtained using the polynomials and data given by *Driscoll* and *Vaughan* [2.86]. *Diels* [2.85] lists n, $dn/d\lambda$ and $d^2n/d\lambda^2$ for materials such as quartz, BK 7, LaSF$_9$, ZnSe and CaF$_2$

for $\lambda \sim 620$ nm. Further information on the dispersion of optical materials can be found in [2.87].

References

2.1   F.P. Schäfer: Principles of dye laser operation, in *Dye Lasers*, ed. by F.P. Schäfer, Topics in Appl. Phys., Vol. 1, 3rd edn. (Springer, Berlin, Heidelberg 1990) pp. 1–89

2.2   F.J. Duarte: Narrow-linewidth pulsed dye laser oscillators, in *Dye Laser Principles*, ed. by F.J. Duarte, L.W. Hillman (Academic, New York 1990) pp. 133–183

2.3   F.J. Duarte: Technology of pulsed dye lasers, in *Dye Laser Principles*, ed. by F.J. Duarte, L.W. Hillman (Academic, New York 1990) pp. 239–285

2.4   B.H. Soffer, B.B. McFarland: Appl. Phys. Lett. **10**, 266 (1967)

2.5   D.J. Bradley, G.M. Gale, M. Moore, P.D. Smith: Phys. Lett. **26A**, 378 (1968)

2.6   G. Magyar, H.J. Schneider-Muntau: Appl. Phys. Lett. **20**, 406 (1972)

2.7   M. Maeda, O. Uchino, T. Okada, Y. Miyazoe: Jap. J. Appl. Phys. **14**, 1975 (1975)

2.8   I. Shoshan, N.N Danon, U.P. Oppenheim: J. Appl. Phys. **48**, 4495 (1977)

2.9   M.G. Littman, H.J. Metcalf: Appl. Opt. **17**, 2224 (1978)

2.10 M.G. Littman: Opt. Lett. **3**, 138 (1978)

2.11 M.G. Littman: Appl. Opt. **23**, 4465 (1984)

2.12 S. Saikan: Appl. Phys. **17**, 41 (1978)

2.13 T.W. Hänsch: Appl. Opt. **11**, 895 (1972)

2.14 E.J. Beiting, K.A. Smith: Opt. Commun. **28**, 355 (1979)

2.15 R. Trebino, J.P. Roller, A.E. Siegman: IEEE J. Quantum Electron. **QE-18**, 1208 (1982)

2.16 S.A. Myers: Opt. Commun. **4**, 187 (1971)

2.17 E.D. Stokes, F.B. Dunning, R.F. Stebbings, G.K. Walters, R.D. Rundel: Opt. Commun. **5**, 267 (1972)

2.18 D.C. Hanna, P.A. Karkkainen, R. Wyatt: Opt. Quantum Electron. **7**, 115 (1975)

2.19 G.K. Klauminzer: U.S. Patent 4 127 828 (Nov. 1978)

2.20 F.J. Duarte, J.A. Piper: Opt. Commun. **35**, 100 (1980)

2.21 A.F. Bernhardt, P. Rasmussen: Appl. Phys. B **26**, 141 (1981)

2.22 F.J. Duarte, J.A. Piper: Appl. Opt. **23**, 1391 (1984)

2.23 F.J. Duarte, R.W. Conrad: Appl. Opt. **26**, 2567 (1987)

2.24 F.J. Duarte, J.A. Piper: Appl. Opt. **20**, 2113 (1981)

2.25 F.J. Duarte, J.J. Ehrlich, W.E. Davenport, T.S. Taylor: Appl. Opt. **29**, 3176 (1990)

2.26 F.J. Duarte, J.A. Piper: Opt. Commun. **43**, 303 (1982)

2.27 F.J. Duarte, J.A. Piper: Am. J. Phys. **51**, 1132 (1983)

2.28 T. Kasuya, T. Suzuki, K. Shimoda: Appl. Phys. **17**, 131 (1978)

2.29 A. Corney, J. Manners, C.E. Webb: Opt. Commun. **31**, 354 (1979)

2.30 M.K. Iles: Appl. Opt. **20**, 985 (1981)

2.31 R. Buffa, S. Cavalieri, M. Matera, M. Mazzoni: Opt. Commun. **58**, 255 (1986)

2.32 P. Dupre: Appl. Opt. **26**, 860 (1987)

2.33 H. Kogelnik: Bell. Syst. Tech. J. **44**, 455 (1965)

2.34 H. Kogelnik: Propagation of laser beams, in *Applied Optics and Optical Engineering*, ed. by R.R. Shannon, J.C. Wyant (Academic, New York 1979) pp. 155–190

2.35 A.E. Siegman: *Lasers* (University Science Books, Mill Valley, CA, 1986)

2.36 A.E. Siegman: J. Opt. Soc. Am. A**2**, 1793 (1985)

2.37 F.J. Duarte: Opt. Quantum Electron. **21**, 47 (1989)

2.38 J. Turunen: Appl. Opt. **25**, 2908 (1986)

2.39 J.P. Tache: Appl. Opt. **26**, 427 (1987)

2.40 D.W. Peters, C.W. Mathews: Appl. Opt. **19**, 4131 (1980)

2.41 F.J. Duarte: IEEE J. Quantum Electron. **QE-19**, 1345 (1983)

2.42 W.W. Morey: Copper vapor laser pumped dye laser, in Proceedings of the International Conference on Lasers '79, ed. by V.J. Corcoran (STS Press, McLean, VA 1980) pp. 365–373
2.43 E. Berik, I. Berik, I. Davidenko: Thermal effects in an excimer-pumped dye laser, in Proceedings of the International Conference on Lasers '88, ed. by R.C. Sze, F.J. Duarte (STS Press, McLean, VA 1989) pp. 397–403
2.44 F.J. Duarte, J.J. Ehrlich, S.P. Patterson, S.D. Russell, J.E. Adams: Appl. Opt. **27**, 843 (1988)
2.45 W. Brouwer: *Matrix Methods in Optical Instrument Design* (W.A. Benjamin, New York 1964)
2.46 O.E. Martinez: IEEE J. Quantum Electron. **QE-24**, 2530 (1988)
2.47 A.G. Kostenbauder: IEEE J. Quantum Electron. **QE-26**, 1148 (1990)
2.48 A.E. Siegman: New developments in laser resonators, in *SPIE Proceedings*, ed. by D.A. Holmes, Vol. 1224 (SPIE, Bellingham, Washington 1990) pp. 2–14
2.49 F.J. Duarte: Opt. Commun. **71**, 1 (1989)
2.50 F.J. Duarte: Opt. Commun. **53**, 259 (1985)
2.51 F.J. Duarte: Appl. Opt. **24**, 1244 (1985)
2.52 F.J. Duarte, J.A. Piper: Opt. Acta **31**, 331 (1984)
2.53 J.R.M. Barr: Opt. Commun. **51**, 41 (1984)
2.54 R. Trebino: Appl. Opt. **24**, 1130 (1985)
2.55 M. Born, E. Wolf: *Principles of Optics* (Pergamon, New York 1975)
2.56 D. Maystre: Integral methods, in *Electromagnetic Theory of Gratings*, ed. by R. Petit, Topics Curr. Phys., Vol. 22 (Springer, Berlin, Heidelberg 1980) pp. 63–100
2.57 W.H. Steel: *Interferometry* (Cambridge University Press, London 1967)
2.58 J. Meaburn: *Detection and Spectrometry of Faint Light* (Reidel, Boston 1976)
2.59 T.J. Pacala, I.S. McDermid, J.B. Laudenslager: Appl. Phys. Lett. **44**, 658 (1984)
2.60 F.J. Duarte, W.E. Davenport, J.J. Ehrlich, T.S. Taylor: Opt. Commun. **84**, 310 (1991)
2.61 E.A. Gavronskaya, A.V. Groznyi, D.I. Staselko, V.L. Strigun: Opt. Spectrosc. **42**, 213 (1977)
2.62 A.V. Aristov, D.A. Kozlovskii, D.I. Staselko, V.L. Strigun: Opt. Spectrosc. **45**, 683 (1978)
2.63 T.J. McKee, J. Lobin, W.A. Young: Appl. Opt. **21**, 725 (1982)
2.64 I.A. McIntyre, M.H. Dunn: Opt. Commun. **50**, 169 (1984)
2.65 L.G. Nair, K. Dasgupta: IEEE J. Quantum Electron. **QE-21**, 1782 (1985)
2.66 E. Berik, B. Davidenko, V. Mihkelsoo, P. Apanasevich, A. Grabchikov, V. Orlovich: Opt. Commun. **56**, 283 (1985)
2.67 M.R. Gorbal, M.I. Savadatti: Pramana J. Phys. **31**, 205 (1988)
2.68 R. Wallenstein, T.W. Hänsch: Opt. Commun. **14**, 353 (1975)
2.69 R. Wallenstein, T.W. Hänsch: Appl. Opt. **13**, 1625 (1974)
2.70 I. Golub, Y. Beaudoin, S.L. Chin: J. Mod. Opt. **36**, 413 (1989)
2.71 F.J. Duarte, R.W. Conrad: Dispersive alexandrite lasers, in Proceedings of the International Conference on Lasers '89, ed. by D.G. Harris, T.M. Shay (STS Press, McLean, VA 1990) pp. 552–554
2.72 F.J. Duarte, J.A. Piper: Appl. Opt. **21**, 2782 (1982)
2.73 R.S. Hargrove, T. Kan: IEEE J. Quantum Electron. **QE-16**, 1108 (1980)
2.74 F.J. Duarte: U.S. Patent 4 822 150 (April, 1989)
2.75 K. Jain: U.S. Patent 4 252 410 (Feb., 1981)
2.76 F.J. Duarte: J. Opt. Soc. Am. A **4**, P(30) (1987)
2.77 F.J. Duarte, D. J. Paine: Quantum mechanical description of N-slit interference phenomena, in Proceedings of the International Conference on Lasers '88, ed. by R.C. Sze, F.J. Duarte (STS Press, McLean, VA 1989) pp. 42–47
2.78 P.A.M. Dirac: *The Principles of Quantum Mechanics*, 4th edn. (Oxford University Press, London 1978)
2.79 R.P. Feynman, R.B. Leighton, M. Sands: *The Feynman Lectures on Physics*, Vol. III (Addison Wesley, Reading, MA 1971)
2.80 F.J. Duarte: Opt. Quantum Electron. **19**, 223 (1987)
2.81 F.J. Duarte: Opt. Quantum Electron. **22**, 467 (1990)
2.82 R.L. Fork, O.E. Martinez, J.P. Gordon: Opt. Lett. **9**, 150 (1984)
2.83 J.D. Kafka, T. Baer: Opt. Lett. **12**, 401 (1987)

2.84 F.J. Duarte: Subtle angular effects in prismatic pulse compression, in Proceedings of the International Conference on Lasers '88, ed. by R.C. Sze, F.J. Duarte (STS Press, McLean, VA 1989) pp. 383–387
2.85 J.C. Diels: Femtosecond dye lasers, in *Dye Laser Principles*, ed. by F.J. Duarte, L.W. Hillman (Academic, New York 1990) pp. 41–132
2.86 W.G. Driscoll, W. Vaughan: *Handbook of Optics* (McGraw-Hill, New York 1978)
2.87 Z. Bor, B. Racz: Appl. Opt. **24**, 3440 (1985)

3. Pulsed Dye Laser Gain Analysis and Amplifier Design

C. Jensen

With 22 Figures

The versatility and utility of organic dye lasers in science and technology continue to increase long after the inception of the dye laser over twenty years ago. Pulsed dye lasers can be configured to generate femtosecond pulses or control chemical processes on a large scale. New laser dyes are rapidly being developed which provide high efficiency and chemical stability for dye lasers at wavelengths ranging from the near-infrared to the ultraviolet. Large scale high average power pulsed dye lasers [3.1–3] are under development in several laboratories for communications, atmospheric propagation studies, isotope separation and many other applications.

The development of a high power pulsed dye laser requires a detailed knowledge of the molecular properties of the laser dye and how those properties are affected by the environment the laser dye experiences within the laser device. For reasons such as operations safety and materials compatibility, laser dyes may be dissolved or suspended in solvents or plastics which may alter the performance of the dye in unpredictable ways. It is therefore essential to have a systematic method for analyzing a laser dye which is simple and accurate yet versatile enough to apply to an extensive range of conditions.

The conventional method of analysis of the molecular and laser gain properties of laser dyes has not changed much since dye lasers were invented [3.4, 5]. The laboratory instrumentation has improved dramatically but the basic method remains tedious and difficult. Unless great care and effort is expended the results cannot be viewed with a high level of confidence. The most important characteristic of a laser dye to be determined is the stimulated emission cross section. The evaluation of this quantity, in the standard method, requires the independent measurement of several properties of the dye including the fluorescence lifetime and the fluorescence quantum yield. The fluorescence quantum yield is the most difficult characteristic of a laser dye to measure absolutely [3.6, 7]. The fluorescence quantum yield of a dye can be measured relative to another dye with a known absolute quantum yield with an accuracy of only about ± 10% [3.8] This error along with the error in the measurement of the lifetime are included in the determination of the stimulated emission cross section.

Considerable effort has been made in synthesizing and analyzing laser dyes which emit at the short wavelengths of the visible spectrum and in the near ultraviolet. New dyes have been analyzed for flashlamp-pumped [3.9–12] and laser-pumped dye lasers. Excimer laser-pumped dyes have received the most

Springer Series in Optical Sciences, Vol. 65
High-Power Dye Lasers   Editor: Francisco J. Duarte
© Springer-Verlag Berlin Heidelberg 1991

study with most of the work directed at XeCl excimer laser pumping [3.13–21]. Nitrogen lasers are a less powerful pump source for dye lasers but are useful for short-wavelength laser dyes [3.22]. The KrF excimer laser is also used to pump laser dyes [3.23–26] to produce tunable ultraviolet laser emission. The photochemical stability of laser dyes exposed to the various pump sources has been studied in detail [3.18, 19, 27–35]. Additional information on the analysis of laser dyes as pertains to flashlamp-pumped dye lasers is provided in Sect. 6.5.5.

A new method for measuring the optical properties of laser dyes has been developed at Los Alamos National Laboratory. This method directly measures the parameters important to laser action and eliminates any need to measure the quantum yield. The principle of this method is to measure the optical saturation properties of the laser dye directly and from these to deduce the optical saturation intensities and excited state absorption cross sections. The saturation analysis method is simpler, more accurate and far less laborious than the conventional method. In addition this method uses standard laboratory equipment, obviating the need to invest in specialized apparatus. The new method has been successfully employed to analyze recently synthesized laser dyes which emit in the near ultraviolet. The results of these analyses have also revealed some unexpected and important processes occurring in the optical dynamics of these dyes. In the following sections of this chapter an analytical technique for measuring the saturation properties of laser dyes is described. The saturation analysis model includes an important excited state absorption not found in the conventional method. The method described herein applies to all laser dyes.

The saturation analysis method for analyzing laser dyes involves the measurement of the optical saturation properties of absorption, fluorescence and stimulated emission. The capabilities of the method to unravel the various competing stimulated processes operating in a laser-pumped dye laser are demonstrated by completely analyzing a generic laser dye. It is pedagogically effective to assign known values to the optical parameters of the dye and evaluate the ability of the analysis method to deduce these values correctly. The laser dye under study is set up to exhibit the competing optical processes which make the analysis of the dye most difficult. Simulated data based on the results of the analysis of one dye (TBS, Lambda Physik and Exciton) is used to demonstrate the method. The absorption band and fluorescence band of TBS are used as the basis for simulated data. The values of the optical parameters assigned to the simulated dye closely resemble those of TBS. The simulated data is analyzed by the same method used to analyze TBS. The use of simulated data allows detailed illumination of the problems associated with pulsed laser analysis by comparing the parameter values, obtained from the analysis, with the correct input parameter values. It is shown that the exact shape of the spatial and temporal profiles of the lasers used in the analysis of laser dyes is crucial to the accurate measurement of the optical parameters.

The experiments associated with the saturation analysis method are performed with dye samples which have low concentrations of laser dye. The use of low concentration dye solutions mitigates several complicating effects such as concentration quenching and amplified spontaneous emission (ASE) [3.36–39]. The effects of ASE on dye laser oscillator performance is covered in Chap. 2.

In Sect. 3.1 the optical dynamics of the energy levels of the laser dye are described for three temporal conditions: short, long and intermediate pulse optical pumping. The short and long temporal pulses refer to time durations compared to the intersystem crossing time which removes population from the first singlet state to a triplet state. The first singlet state produces laser action whereas the triplet state derogates from laser action [3.40–43]. The intersystem crossing time [3.44] in useful laser dyes can range from a few nanoseconds to milliseconds. In this case the quality of a dye is determined by the relative rates of intersystem crossing and fluorescence and the duration of the pump radiation. Temporal pulses of a nanosecond or less in duration are short pulses and temporal pulses a few hundred nanoseconds or longer are long pulses. The duration of intermediate temporal pulses emitted by common pump laser sources typically range from a few nanoseconds to a few tens of nanoseconds. The effect of intersystem crossing and triplet state build-up in these different time regimes is graphically illustrated and the validity of the steady state approximation for simplifying the optical dynamics equations is quantified.

The steady state theory for the intermediate pulse optical pumping case is developed for the commonly used pulsed dye laser pump sources, N_2 lasers, Nd:YAG lasers and excimer lasers and is presented in Sect. 3.2. The theory will include excited state absorptions at the dye laser wavelength. The important saturation properties of the laser dye are defined such that once they are measured they will completely determine all of the relevant optical parameters.

In Sect. 3.3 the experiments and analysis are described that determine the saturation properties of the laser dye. The sensitivity of the analysis to the measurement accuracy of the spatial and temporal profiles of pump laser and dye laser pulses is quantified. The numerical data reduction techniques needed to analyze the data in Sect. 3.3 are derived in Sect. 3.4.

A procedure for the simple and accurate determination of the quantum yield for fluorescence of the laser dye from its saturation properties is presented in Sect. 3.5. The self-consistency and completeness of this method for analyzing laser dyes is demonstrated with the correct recovery of the value of the intersystem crossing rate assumed initially in the simulated data.

A simple model for designing a dye laser amplifier based on the information learned from the analysis of the laser dye is described in Sect. 3.6. The usual constraints that are imposed on an amplifier design are the damage threshold for the amplifier optics (Chap. 4) and the desired amplifier gain. This procedure provides the optimization of the amplifier performance subject to the physical constraints and limitations of the optics. The effects of excited state absorption in the laser dye on the amplifier design are also included. The correspondence between the nomenclature used in Chap. 6 and that used in Chap. 3, for the optical parameters of a laser dye, is given in Appendix A of Chap. 6.

3.1 Time Dependent Dynamics of a Laser Dye

The molecular energy levels [3.4, 5] of an organic dye are shown in Fig. 3.1. The unperturbed dye molecules reside in the ground electronic state, S_0. In the case of a laser-pumped dye laser, the pump laser radiation is absorbed by the dye molecules in the ground state with a cross section σ_{01}. These molecules are promoted to a vibronic level in the first excited singlet state, S_1, which thermally relaxes to the lowest vibronic level of S_1 on a picosecond timescale. From the lowest vibronic energy level of S_1, several processes can occur. The pump radiation can be absorbed again, this time with a cross section, σ_{12}, and the population moves to a higher lying singlet electronic state, S_n. The population in S_n thermally relaxes back to S_1 also on a picosecond timescale. The population in S_1 will also fluoresce back to a vibronic level of S_0, where it then thermally relaxes to the ground state. The fluorescence lifetime is typically one to eight nanoseconds. If probe laser radiation with a wavelength within the fluorescence band is also present then the population in S_1 will radiate by stimulated emission with a cross section σ_e or absorb radiation with a cross section σ_{12}^L. Absorption of the probe radiation can occur from S_0 with a cross section σ_{01}^L.

Population can also be lost from S_1 by the non-radiative process of intersystem crossing to the lowest triplet electronic state, T_1, with a rate constant, k_{ST} [3.45]. Once a population is in T_1 it can absorb both the pump and probe radiation. The radiation absorbed by the triplet state population is converted into heat by thermal relaxation of the excited triplet state, T_2. The population in T_1 experiences a very slow decay to the ground state by thermal relaxation or

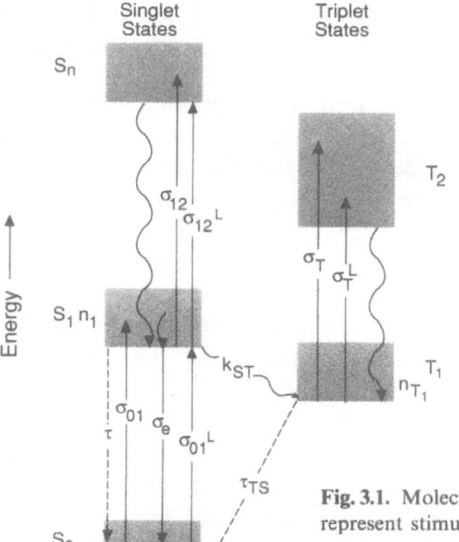

Fig. 3.1. Molecular energy states of an organic dye. Solid lines represent stimulated optical transitions, dashed lines represent spontaneous optical transitions and wavy lines refer to radiationless transitions

phosphorescence with a lifetime, τ_{TS}. The strength of the $T_2 \leftarrow T_1$ absorption can be measured [3.12, 46–48] using the method of *McClure* [3.49]. The use of triplet quenchers in the dye solution can lower the value of τ_{TS} and improve the performance of the dye laser [3.17, 50–52].

The only processes which ultimately contribute to the amplification of the probe laser are 1) the ground state absorption of the pump beam which provides the energy for amplification and 2) stimulated emission which provides the process of amplification. All other optical processes discussed above are losses to the efficiency of the dye laser amplifier.

A more detailed molecular state model for the processes induced by pump laser radiation absorption is shown in Fig. 3.2. The equations which represent the time rate of change of the energy levels are given below:

$$\frac{dn_0}{dt} = -n_0 \sigma_{01} I_P + \frac{n'_0}{\tau_0} - \frac{n_0}{\tau_u} + \frac{n_{T_1}}{\tau_{TS}} \tag{3.1}$$

$$\frac{dn'_0}{dt} = \frac{n_1}{\tau_{rad}} + \frac{n_0}{\tau_u} - \frac{n'_0}{\tau_0} + n_1 k_{IC} \tag{3.2}$$

$$\frac{dn_1}{dt} = \frac{n'_1}{\tau_1} - \frac{n_1}{\tau_{rad}} - n_1 \sigma_{12} I_P - n_1 k_{IC} - n_1 k_{ST} \tag{3.3}$$

$$\frac{dn'_1}{dt} = n_0 \sigma_{01} I_P + \frac{n'_2}{\tau_2} - \frac{n_1}{\tau_1} \tag{3.4}$$

$$\frac{dn'_2}{dt} = n_2 \sigma_{12} I_P - \frac{n'_2}{\tau_2} \tag{3.5}$$

$$\frac{dn_{T_1}}{dt} = n_1 k_{ST} - n_{T_1} \sigma_T I_P + \frac{n_{T_2}}{\tau_T} - \frac{n_{T_1}}{\tau_{TS}} \tag{3.6}$$

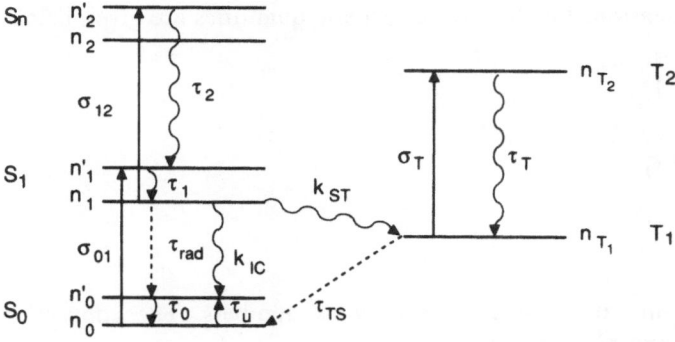

Fig. 3.2. Molecular energy levels and optical transitions affected by pump laser radiation. The population of a level is n_i

$$\frac{dn_{T_2}}{dt} = n_{T_1}\sigma_T I_P - \frac{n_{T_2}}{\tau_T} . \tag{3.7}$$

Each of these equations represents the time rate of change of the population of one of the energy levels which is affected by the pump laser radiation. Equation (3.1) describes the time rate of change of the ground state population, n_0, which loses population by absorption of the pump radiation with an absorption cross section σ_{01}. The pump radiation intensity is I_P. In the second term, the population, n_0, is increased by the thermal relaxation (with lifetime τ_0) from the energy level which is the lower level of the fluorescence transition. The third term is the rate of filling population n_0' from the ground state with lifetime τ_u to maintain the equilibrium Boltzmann distribution. The population of the ground state is also increased by the population transfer from the lowest triplet level with lifetime τ_{TS}. Equation (3.2) denotes the rate of change of the excited vibrational state of the ground electronic state S_0. The population of this state is increased by fluorescence (τ) and radiationless internal conversion (k_{IC}) and strives towards thermal equilibrium with thermal relaxation lifetimes τ_0 and τ_u. The relative values of τ_0 and τ_u are given by $n_0'/n_0 = \tau_u/\tau_0$ where the populations are taken to be their equilibrium values. Equation (3.3) shows the time dependance of the lowest vibrational state of the first excited singlet state S_1. This energy level is populated by thermal relaxation from the excited vibrational level (n_1') which is directly populated by absorption of the pump radiation (I_P). The population n_1 experiences several processes including fluorescence down to S_0, absorption of pump radiation to S_n, internal conversion to S_0 and intersystem crossing to the triplet state T_1. Equations (3.4) and (3.5) involve both the ground state and excited state absorptions and the thermal relaxation to the n_1 population of the S_1 state. The kinetics of the triplet states are described by (3.6) and (3.7). The population n_{T_1} in T_1 results from radiationless intersystem crossing from S_1. Once in T_1 the triplet population experiences absorption to and thermal relaxation from T_2 and eventually transfer back to the ground state with lifetime τ_{TS}, as described by (3.7). The population transfer to the ground state can proceed by phosphorescence, intermolecular quenching or thermal relaxation. The dimensions for the various generic quantities are given below:

n_i (particles/cm^3)

σ_i(cm^2)

I_i (photons/cm^2 s)

τ_i (s)

k_{ij} (s^{-1}) .

Particles and photons are considered simply as numbers. The conversion between Watt/cm^2 and photons/s cm^2 is given by

$$1 \text{ W/cm}^2 = \lambda(\text{nm})/1.988 \times 10^{-16} \text{ photons/s cm}^2 .$$

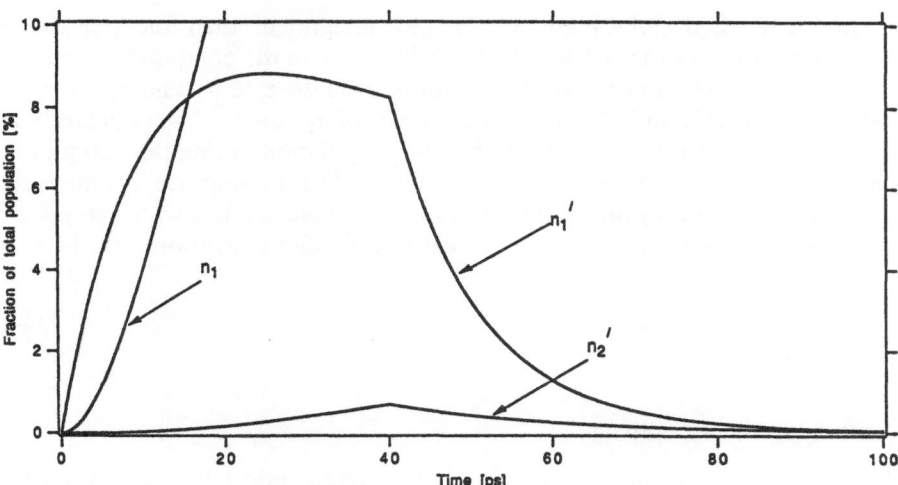

Fig. 3.3. Time dependence of energy level populations during and after a 30 ps (FWHM) pump laser pulse with temporal peak occurring at 40 ps

The photon flux equivalent in intensity to 1 Watt/cm^2 at the XeCl excimer pump laser wavelength (308 nm) is 1.55×10^{18} photons/cm^2 s.

The dynamics of the energy level populations in Fig. 3.3 are demonstrated by assigning a temporal profile to the pump intensity I_P and numerically integrating (3.1–7). The various population densities, n_i, and the pump intensity I_P will have a temporal dependence whereas the rest of the quantities in (3.1–7) will be constants. Values are assigned to the constants which would be found in an excimer pumped near-UV laser dye. The singlet state thermal relaxation lifetimes have been increased from the actual 1 ps to 10 or 20 ps to enhance graphic representation.

λ(pump) = 308 nm	λ(emission) = 400 nm
$\sigma_{01} = 2.5 \times 10^{-16}$ cm^2	$\sigma_{12} = 1.0 \times 10^{-16}$ cm^2
$\tau_0 = 10$ ps	$\tau_1 = 10$ ps
$\tau_2 = 20$ ps	$\tau_T = 20$ ps
$\tau_u = 10$ ns	$\tau_{rad} = 1.176$ ns
$k_{ST} = 1.5 \times 10^8$ s^{-1}	$\tau_{TS} = 1.0$ μs

The initial conditions at $t = 0$ are $n_0(0) = 1.0 \times 10^{16}$ cm^{-3} and $n_0'(0) = 1.0 \times 10^{13}$ cm^{-3}.

3.1.1 Short Pulse Pumping

When a temporal profile for the pump intensity is taken to be a Gaussian pulseshape with a pulsewidth (FWHM) of 30 ps and a peak intensity of

30 MW/cm², and (3.1–7) are numerically integrated, then the population dynamics shown in Fig. 3.3 are obtained. The peak of the pump pulse occurs at 40 ps. The competition between the rate of excitation to S_1 and S_n from the pump pulse is evident in the temporal profiles of n_1' and n_2'. The population n_1 achieves a peak level of $> 60\%$ of the total population during the pump pulse then decays to S_0 with a fluorescence lifetime τ. The fluorescence lifetime is the observed optical decay time from a fluorescing sample pumped with a very short pulse. The fluorescence lifetime is related to the other relaxation rates by

$$\frac{1}{\tau} = \frac{1}{\tau_{\mathrm{rad}}} + k_{\mathrm{ST}} + k_{\mathrm{IC}} \ . \tag{3.8}$$

3.1.2 Long Pulse Pumping

We now consider the opposite extreme in the pulsed optical dynamics where the temporal profile of the pump intensity is long compared to all the relaxation time scales of the dye molecule. A temporal profile representing the output of a flashlamp is shown in Fig. 3.4a. The effects of the triplet states on the population n_1 of the fluorescing energy level in S_1 can be illustrated by numerically propagating the temporal pulse through (3.1–7). The same values for the optical constants as above are assumed except for τ_{TS} which is varied. The peak intensity for the temporal profile is 1.0 MW/cm². The results of the calculation of n_1 and n_{T_1} as a function of time for two different values of $k_{\mathrm{TS}} = 1/\tau_{\mathrm{TS}}$ are shown in Figs. 3.4b,c. In Fig. 3.4b where $k_{\mathrm{TS}} = 5.0 \times 10^7 \ \mathrm{s}^{-1}$ the calculation shows that most of the excited population ends up in T_1 with very little in S_1, which means that under these conditions the dye sample will act as an absorber ($T_2 \leftarrow T_1$ transitions) rather than as an amplifier ($S_0 \leftarrow S_1$ transitions). By increasing the $S_0 \leftarrow T_1$ relaxation rate by a factor of ten as shown in Fig. 3.4c the situation becomes reversed. Here the singlet state population n_1 is in preponderance over the triplet state population n_{T_1} and the dye sample can act as an amplifier of electromagnetic radiation present at the dye emission wavelength. The value of k_{TS} can be experimentally varied by adding triplet quenchers such as cyclooctatetraene and O_2 to the dye solution [3.17, 49–51].

3.1.3 Excimer Laser Pumping

The optical dynamics of laser-pumped dyes fall into an intermediate case where the pulsewidth of the temporal profile of the pump intensity is comparable to the intersystem crossing time. The temporal profile of a laser pulse from an XeCl excimer laser (Lambda Physik 203, neon-rich mixture) is shown in Fig. 3.5a. Taking the same values for the optical constants as before, with $k_{\mathrm{ST}} = 1.5 \times 10^8 \ \mathrm{s}^{-1}$, a peak pump laser intensity of 1.0 MW/cm² and integrating (3.1–7) over this temporal profile, the curves for the time dependence of n_1 and n_{T_1} obtained are shown in Fig. 3.5b. The time dependence of the fluorescing level

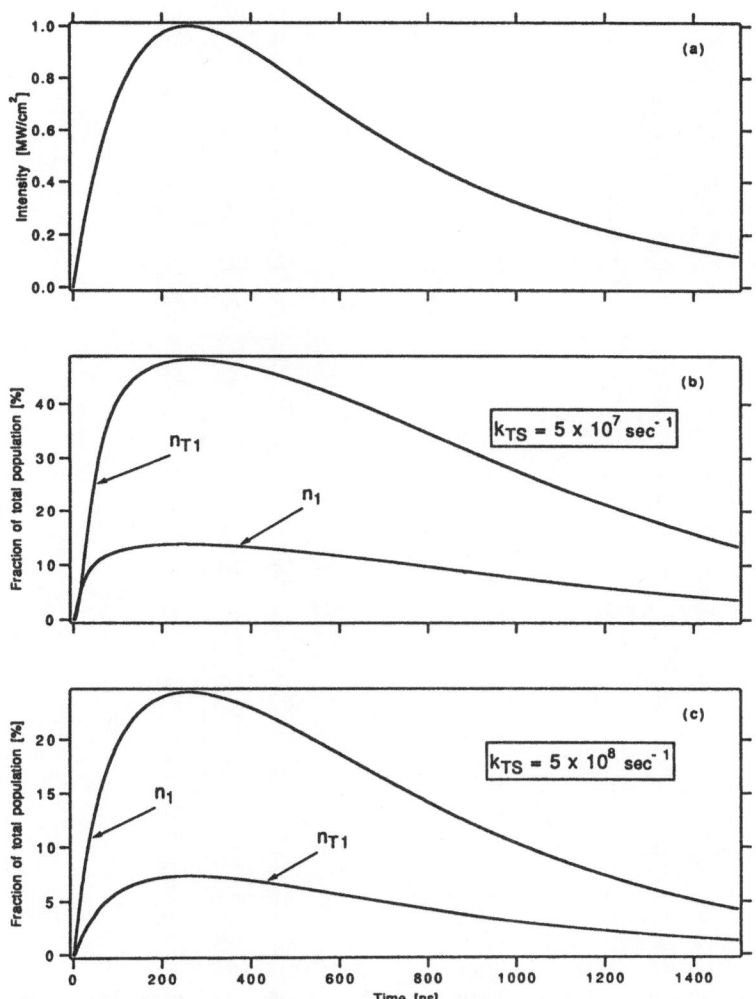

Fig. 3.4. Population dynamics of energy levels responding to a flashlamp pump pulse; (a) temporal profile of flashlamp pulse, (b) and (c) time dependence of singlet and triplet level populations

population n_1 replicates the temporal profile of the pump intensity. This indicates that the pump pulse and the singlet states are in dynamic steady state. Hence, the very fast thermal relaxation processes (Sect. 3.1.1) can be ignored.

Figure 3.5a also shows the experimental temporal profile of a laser pulse from an excimer-pumped dye laser oscillator. The laser dye in the oscillator (TBS) has optical constants very similar to the ones assumed for these calculations. The shape of the temporal profile for the probe pulse in Fig. 3.5a also imitates the pump pulse temporal profile within the limits of dye laser threshold,

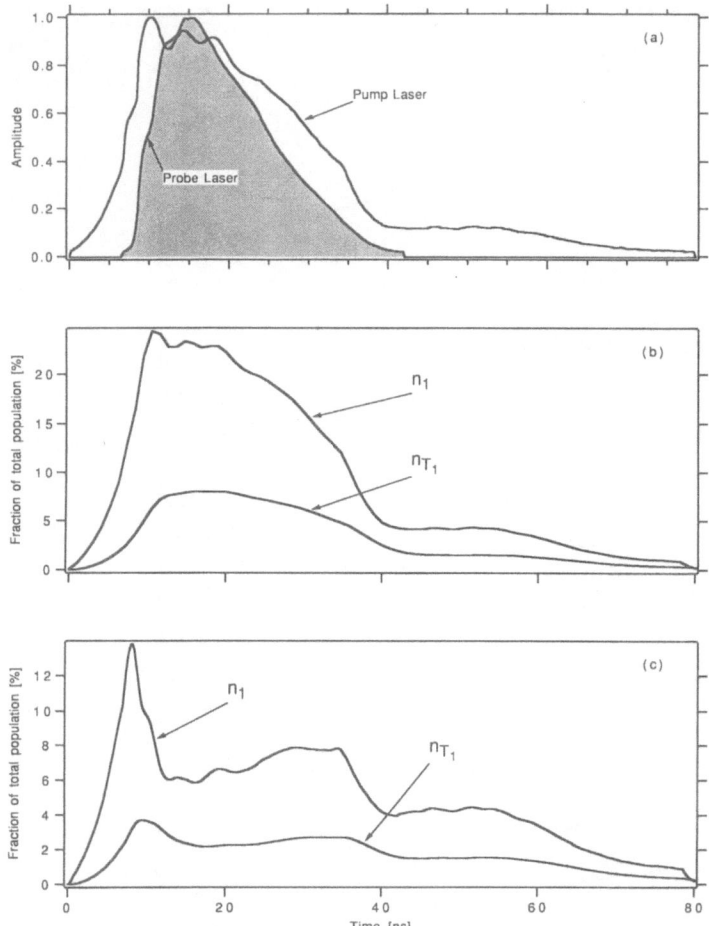

Fig. 3.5. Population dynamics of energy levels responding to pump and probe laser pulses; (a) temporal profiles of excimer (pump) laser and excimer-pumped dye (probe) laser, (b) and (c) time dependence of singlet and triplet level populations

which implies that the values assumed for the optical constants are reasonable. The effects of probe radiation on the singlet and triplet state dynamics is seen by adding terms for stimulated emission to (3.1–7) and time-integrating over both the pump and probe temporal profiles. To include stimulated emission in (3.1–7) the term $n_1 \sigma_e I$ is added and subtracted from (3.3) and (3.4) respectively. The stimulated emission cross section σ_e at the probe laser wavelength, 400 nm, is assigned a value of 2.34×10^{-16} cm^2 (which is the value used in later sections). The probe pulse intensity I peak value is 10 MW/cm^2. No spatial dependence of the pump and probe intensities is included. The calculation yields the curves in

Fig. 3.5c where a significantly reduced population is seen in both S_1 and T_{T_1} owing to the rapid depopulation of S_1 by the emission induced by the probe intensity. Hence, under the normal operating conditions of an excimer-pumped dye laser amplifier where both pump and probe radiations are present, the population in S_1 is not sufficient to allow significant filling of T_1. If k_{TS} is rendered large enough (and τ_{TS} small enough) by appropriate application of triplet quenchers then whatever population does enter T_1 returns to S_0 in time to be recycled through the singlet states during the pump and probe pulses. The conclusion from this result is that the dynamics of the singlet states are in steady state with the excimer laser pump pulse and the effects from the triplet states can be ignored.

3.2 Steady State Approximate Theory

The examples given in Sect. 3.1 illustrate the molecular state population dynamics of a laser dye under different pumping conditions and the effect of probe laser radiation in the case of an excimer laser-pumped laser dye. It is not necessary to retain the details of the results provided by (3.1–7) for an excimer laser-pumped dye. It is useful to have an approximate simpler theory for predicting dye laser behavior. This is done by assuming that the non-radiative relaxation processes do not contribute to the performance of the dye laser (Sect. 3.1.3) and solving the equations for the singlet state populations in the steady state limit. With the assumption of the steady state approximation the energy level diagram for the dye becomes Fig. 3.6.

The operational characteristics of a pulsed dye laser amplifier can be calculated by solving the photon propagation equations for the pump and probe intensities. The usual experimental configuration of the pump and probe beams is shown in Fig. 3.7. The pump laser beam, typically from an excimer, Nd:YAG or N_2 laser, is cylindrically focused along the length of a side of the dye amplifier to provide high intensity over a length of a few centimeters. The probe beam propagates through the dye amplifier in a direction perpendicular to and intersecting the pump beam. The pump beam intensity provides the excitation to S_1 from which emission transitions are stimulated by the probe beam intensity, thereby increasing the probe beam intensity. The propagation of the

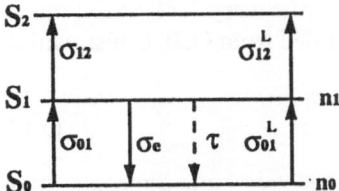

Fig. 3.6. Simplified molecular energy level diagram for steady state theory

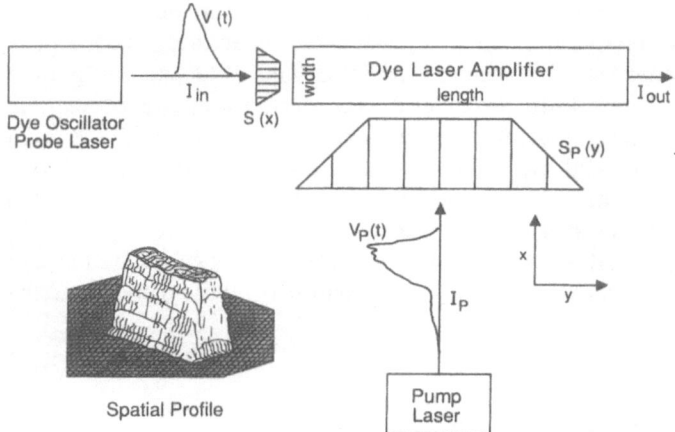

Fig. 3.7. Block diagram (top view) of dye laser amplifier, probe laser and pump laser. Temporal and digitized spatial amplitude profiles of pump and probe lasers used in calculations are shown in approximate form. The three-dimensional spatial profile is of an homogenized excimer laser beam (intensity is in the vertical direction) .

pump beam intensity in the x direction is described by

$$\frac{dI_P}{dx} = - n_0\sigma_{01}I_P - n_1\sigma_{12}I_P \tag{3.9}$$

and the propagation of the probe beam intensity in the y direction is given by

$$\frac{dI}{dy} = n_1\sigma_e I - n_0\sigma_{01}^L I - n_1\sigma_{12}^L I \tag{3.10}$$

where σ_{01}^L and σ_{12}^L are the $S_1 \leftarrow S_0$ and $S_2 \leftarrow S_1$ absorption cross sections at the probe laser wavelength respectively. In Sect. 3.1.3 it was demonstrated that the triplet states do not participate appreciably under pulsed amplifier conditions and n_1 maintains a steady state with the applied intensities. The time rate of change of n_1 is given by

$$\frac{dn_1}{dt} = n_0\sigma_{01}I_P + n_0\sigma_{01}^L I - n_1\sigma_e I - \frac{n_1}{\tau} \tag{3.11}$$

where the terms involving the triplet states and the very fast radiationless transitions among the singlet states have been dropped from (3.3). Conservation of particles is observed by

$$n = n_0 + n_1 \tag{3.12}$$

where n is the total density of dye molecules. The steady state approximation (justified in Sect. 3.1.3) sets

$$\frac{dn_1}{dt} = 0 \ . \tag{3.13}$$

Solving (3.11–13) for n_0 and n_1 in terms of n yields

$$n_0 = \frac{n(1 + \sigma_e \tau I)}{1 + \sigma_{01} \tau I_P + \sigma_e \tau I + \sigma_{01}^L \tau I} \tag{3.14}$$

and

$$n_1 = \frac{n(\sigma_{01} \tau I_P + \sigma_{01}^L \tau I)}{1 + \sigma_{01} \tau I_P + \sigma_e \tau I + \sigma_{01}^L \tau I} \ . \tag{3.15}$$

The saturation intensities for pump laser absorption, stimulated emission and probe laser absorption are defined as

$$I_{PS} = (\sigma_{01} \tau)^{-1} \ , \tag{3.16a}$$

$$I_S = (\sigma_e \tau)^{-1} \tag{3.16b}$$

and

$$I_{LS} = (\sigma_{01}^L \tau)^{-1} \ , \tag{3.16c}$$

respectively. An applied intensity equal to the saturation intensity corresponds to a photon flux such that there is one photon passing through one cross section (area) per fluorescent lifetime (τ) [3.53]. By substituting (3.16) into (3.14, 15) the singlet state densities can be expressed in terms of the total density and saturation intensities

$$n_0 = \frac{n(1 + I/I_S)}{1 + I_P/I_{PS} + I/I_S + I/I_{LS}} \tag{3.17}$$

$$n_1 = \frac{n(I_P/I_{PS} + I/I_{LS})}{1 + I_P/I_{PS} + I/I_S + I/I_{LS}} \ . \tag{3.18}$$

Substituting (3.14–16) into (3.9) and (3.10) the photon propagation equations are obtained:

$$\frac{dI_P}{dx} = \frac{-n\sigma_{01} I_P [1 + I/I_S + R(I_P/I_{PS} + I/I_{LS})]}{1 + I_P/I_{PS} + I/I_S + I/I_{LS}} \tag{3.19}$$

$$\frac{dI}{dy} = \frac{n\sigma_{01} I_P [(1 - R_e)I/I_S] - n\sigma_{01}^L I(1 + R_e I/I_S)}{1 + I_P/I_{PS} + I/I_S + I/I_{LS}} \ , \tag{3.20}$$

where

$$R = \frac{\sigma_{12}}{\sigma_{01}} \tag{3.21a}$$

$$R_e = \frac{\sigma_{12}^L}{\sigma_e} \, . \tag{3.21b}$$

With the values of all the optical constants in (3.19, 20) known, the performance of a side-pumped pulsed dye laser amplifier can be modeled.

3.3 Experimental Series for Analyzing a Laser Dye

The task for the experimentalist is to measure the values of the optical constants in (3.19, 20). The optical constants are:

$$\sigma_{01}, \sigma_{01}^L, I_{PS}, I_S, I_{LS}, R \quad \text{and} \quad R_e \, .$$

In the conventional analysis of laser dyes σ_{01} and σ_{01}^L are calculated from an absorption spectrum of the dye. The fluorescence lifetime τ is measured using a picosecond laser and a fast detector such as a streak camera. The saturation intensities I_{PS} and I_{LS} are then calculated from (3.16a, c). The value of R is determined from measuring the transmission of the pump beam through a dye sample as a function of pump intensity [3.5]. The most difficult optical constant to determine is σ_e. In addition to the fluorescence band and fluorescence lifetime, it requires the measurement of the quantum yield which is the most difficult (and inaccurate) characteristic of a laser dye to measure [3.6–8]. Once σ_e is obtained then the saturation intensity, I_S, is calculated with (3.16b). The value of σ_{12}^L is also extremely difficult to measure and requires independent knowledge of σ_e [3.54]. The stimulated emission strength of a dye laser gain medium is reduced by σ_{12}^L which makes σ_{12}^L a very important quantity. When σ_{12}^L and σ_e are known then R_e is computed from (3.21b). The complete evaluation of a laser dye by these methods is quite laborious and difficult but the results are necessary for accurate modeling of the dye laser.

It would be a tremendous advantage to be able to determine the values of I_{PS}, I_S, I_{LS}, R and R_e without the need for measuring the fluorescence lifetime or the quantum yield.

A series of experiments can be devised which measure the values of the optical constants I_{PS}, I_S, I_{LS}, R and R_e directly without measuring the fluorescence lifetime or quantum yield. In fact, the fluorescence lifetime and quantum yield can be determined from the optical constants once these optical constants have been measured. The advantage of the method of direct measurement of the optical constants is that it requires only the equipment needed to measure the temporal and spatial profiles of the laser pulse and the pulse energy. The required equipment is inexpensive and standard in most laboratories employing dye lasers. The experiments needed to measure these constants are simpler and much more accurate than those employed in quantum yield measurements and therefore the determination of σ_e and σ_{12}^L is more accurate.

The value of I_{PS} can be deduced from a measurement of the relative fluorescence yield of the dye as a function of the pump laser intensity. With this result and the absorption spectrum, $I_{LS} = I_{PS} \times \sigma_{01}/\sigma_{01}^{L}$ can be calculated. The values of I_{PS} and R can both be obtained from the measurement of pump laser transmission through a dye sample as a function of pump laser intensity [3.55]. The measurement of gain in a dye laser amplifier provides the values of I_S and R_e. A summary of the experimental series is listed in Table 3.1.

To demonstrate the efficacy of the experimental series at determining all the values of the optical constants needed to evaluate a laser dye, values for the optical constants are assigned and used to calculate simulated experimental data. The data are generated by numerically integrating the coupled photon propagation differential equations (3.19, 20). Since (3.19) and (3.20) are the equations to which real data are fit, the only difference between the simulated data and real data is the lack of experimental error. As such, the data is in a sense perfect so the fit of the data to the same equations should be exact. The calculations utilize real experimental temporal and spatial profiles for the laser pulses. The values of the optical constants and the absorption and fluorescence spectra closely resemble a near-UV laser dye, TBS. From a fitting routine, described in Sect. 3.4, the values of the optical constants can be extracted from the data. The use of simulated data provides an absolute comparison of the data analysis results with the true values of the optical constants and allows exploration of the level of experimental detail required in the analysis for accurate

Table 3.1. Experimental Series

No.	Parameters(s)	Technique
1.	σ_{01}, σ_{01}^{L}	Absorption spectrum with a standard spectrometer
2.	I_{PS}	Relative fluorescence yield with an optical multichannel analyzer
3.	I_{PS}, R	Transmission vs. I_P through a dye sample
4.	I_S, R_e	Gain vs. I through an amplifier at a fixed I_P

Table 3.2. Optical Parameters of a Laser Dye

$\lambda(\text{pump}) = 308$ nm	$\lambda(\text{emission}) = 400$ nm
$\sigma_{01} = 2.50 \times 10^{-16}$ cm^2	$\sigma_e = 2.34 \times 10^{-16}$ cm^2
$\sigma_{12} = 1.00 \times 10^{-16}$ cm^2	$\sigma_{12}^{L} = 0.47 \times 10^{-16}$ cm^2
$\tau = 1.00$ ns	$\sigma_{01}^{L} = 1.50 \times 10^{-19}$ cm^2
$R = 0.40$	$R_e = 0.20$
$I_{PS} = 2.58$ MW/cm^2	$I_S = 2.12$ MW/cm^2
$k_{ST} = 1.5 \times 10^{8}$ s^{-1}	$I_{LS} = 3310$ MW/cm^2

results. As is demonstrated in Sects. 3.3.1, 2 some commonly used approximations for estimating the temporal pulse width and beam area of pulsed lasers introduce significant errors into the fitting of experimental (or simulated) data.

The optical constants for the simulated dye are listed in Table 3.2. These are the constants that are to be determined experimentally. The optical constants in Table 3.2 are consistent with the values used in (3.1–7) to generate Figs. 3.3–5.

3.3.1 Fluorescence Yield

Starting with item (2) in Table 3.1, the relationship between the fluorescence yield and the pump laser intensity can be derived from the equation for fluorescence rate. Combining (3.11), (3.12) and (3.17) and setting $I = 0$, the expression for the fluorescence rate is

$$\frac{n_1}{\tau} = \frac{n\sigma_{01}I_P}{1 + I_P/I_{PS}} . \tag{3.22}$$

The fluorescence yield is measured by irradiating a thin sample of weak dye solution with the pump laser and collecting the fluorescence emitted from the sample at a non-specularly reflected angle to the pump laser beam direction. The fluorescence is collected with a lens or optical fiber and directed into a monochromator. The dispersed fluorescence from the monochromator is detected and temporally integrated by an optical multi-channel analyzer (OMA). The pump intensity is varied and at each intensity the fluorescence spectrum is taken by the OMA. The integral of the fluorescence spectrum over wavelength is taken to be the relative fluorescence yield. A commercial OMA will calculate the integral automatically. If an OMA is unavailable the amplitude from the total fluorescence measured with a photodiode detector can be used as the relative fluorescence yield.

Figure 3.8 shows the absorption and fluorescence spectra of the dye to be evaluated as measured with a standard spectrometer and an OMA, respectively. The relative fluorescence yield, FY, is related to pump intensity by

$$FY = \frac{AI_P}{1 + BI_P} , \tag{3.23}$$

where A is a scaling factor and $B = 1/I_{PS}$. Simulated fluorescence yield data is generated by numerically propagating the pump laser pulse in Fig. 3.5a through (3.1–7), numerically integrating over time, and counting the emission photons. The peak intensity of the pump laser pulse is varied from zero to 40 MW/cm^2. The fluorescence yield data is displayed as boxes in Fig. 3.9. The ordinate axis is in arbitrary units as would be recorded in an experiment. A least-squares fit to the parameters A and B in (3.23) is shown as a solid line in Fig. 3.9. The value obtained for B corresponds to $I_{PS} = 5.84$ MW/cm^2 which is nearly a factor of two greater than the correct value listed in Table 3.2. The reason for this

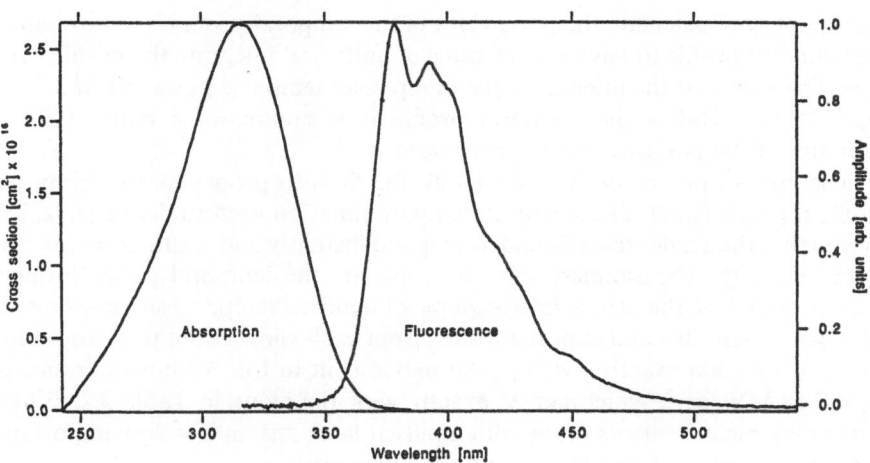

Fig. 3.8. Absorption and fluorescence bands of laser dye. Left ordinate is for the absorption band and right ordinate is for the fluorescence band

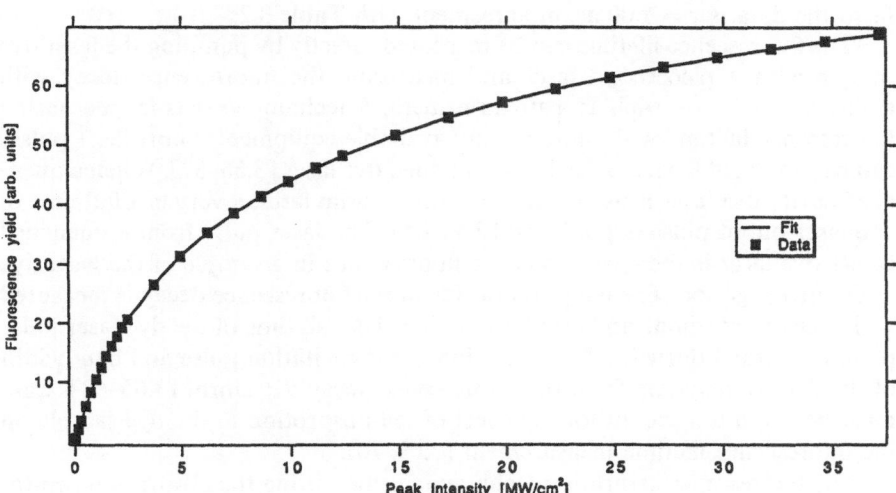

Fig. 3.9. Relative fluorescence yield from laser dye vs. pump laser intensity

discrepancy is an important one to identify, for it is critical for the accurate application of these techniques and to pulsed lasers in general. The value of I_P at each data point in the fit to (3.23) was calculated from the pump laser pulse energy, E_P (Joules), the beam area, a, and the pulsewidth (FWHM), Δt_P (~ 23 ns), of the temporal profile in Fig. 3.5a

$$I_P = \frac{E_P}{a \Delta t_P} \ . \tag{3.24}$$

Another way to calculate the pulsewidth of the temporal profile is to normalize the temporal profile to have a peak value of unity and integrate the profile over time. The value for the integral of the pump laser temporal pulse in Fig. 3.5a is $\Delta t_P = 25.7$ ns. Unless the temporal profile is a square wave both of these estimates of the intensity will be inaccurate.

The correct procedure for calculating the fit is to propagate the temporal profile through (3.23). The steady state approximation used to derive (3.22, 23) implies that the single states S_0 and S_1 respond instantly and continuously to the pump intensity. The simplest way to propagate the temporal pulse through (3.23) is to digitize the profile into a couple of hundred temporal slices, calculate a FY with each slice and sum the results from each slice. Using this procedure the fit to the data exactly overlaps the previous fit in Fig. 3.9 but the result is $I_{PS} = 2.58$ MW/cm^2, which agrees exactly with the value in Table 3.2. When performing measurements on or with a pulsed laser system it is very important to treat the temporal profile explicitly and accurately.

With I_{PS} known from relative fluorescence yield vs. pump laser intensity experiment and σ_{01} known from the absorption spectrum, the fluorescence lifetime can then be calculated from (3.16a). The value calculated from the latter fit to the data is $\tau = 1.00$ ns, in agreement with Table 3.2.

The fluorescence lifetime can be measured directly by pumping the laser dye sample with a picosecond laser and measuring the fluorescence decay with a very fast detector, such as a streak camera. A technique exists for measuring fluorescence lifetime with more readily available equipment. Short laser pulses can be generated from a pulsed laser-pumped dye laser [3.56, 57]. When a quenched cavity dye laser is used with a standard pump laser, a very fast fall time of dye laser ouput pulse is produced [3.58–64]. The laser pulse from a quenched cavity dye laser is then used to excite fluorescence in a sample of the laser dye under investigation. The temporal profile of the fluorescence decay is measured with a fast photodiode and oscilloscope. The fast fall time of the dye laser pulse permits a straightforward deconvolution of the excitation pulse and bandwidth of the detection system from the fluorescence decay waveform [3.65–68]. Care must be taken to account for the effect of self-absorption in the dye sample on the fluorescence lifetime measurement [3.69, 70].

The fluorescence saturation technique for measuring the absorption saturation intensity and determining the fluorescence lifetime is easier and more accurate for short lifetime laser dyes (< 3 ns) than the direct measurement method. At extremely high pump intensities excited state absorption can quench fluorescence [3.71]; however these intensities are not required for the fluorescence saturation technique.

3.3.2 Transmission

The transmission properties of laser dyes [3.55, 72–74] have been widely used to measure excited state absorption [3.75–90] and to evaluate dyes as saturable absorbers [3.91–97] for temporally shaping laser pulses. The measurement of transmission as a function of pump laser intensity through a dye sample quantifies the saturation and excited state absorption properties of a laser dye. Transmission measurements can also reveal parasitic nonlinear absorption processes occurring in the host material.

Item (3) in Table 3.1 in the experimental series for evaluating the laser dye is to measure transmission of the pump laser beam through a dye sample as a function of pump laser intensity. The optical constants which can be determined from the results of this experiment are I_{PS} and the ratio of absorption cross sections at the pump laser wavelength R. Obviously, if I_{PS} is already known from the FY vs. I_P (Sect. 3.3.1) experiment then the fit to the transmission data need only be for R. The transmission data can however provide a cross check to the fluorescence yield result for I_{PS}, or if a fluorescence detector is unavailable then the transmission data can be used to determine both parameters. The equation for the propagation of the pump laser intensity through a dye sample is obtained from (3.19) by setting the probe intensity $I = 0$

$$\frac{dI_P}{dx} = -\frac{n\sigma_{01}I_P(1 + RI_P/I_{PS})}{1 + I_P/I_{PS}} . \tag{3.25}$$

The solution of this differential equation was first provided by *Teschke* et al. [3.5]. Since terms will later be added to (3.25) which will render it insoluble, and with the need to use explicit temporal and spatial profiles, the fitting routine (described iṅ Sect. 3.4) will be applied directly to (3.25). Transmission data was generated assuming a path length through the dye sample of 0.4 cm, a dye density of 1.0×10^{16} cm^{-3} (1.7×10^{-5} M) and numerically integrating (3.25) over temporal and spatial profiles. The spatial profile of the pump laser beam is taken to be a frustum of a pyramid, as would be obtained from a commercial beam homogenizer. An example of such a spatial profile is shown in Fig. 3.7. The width and length at the top of the spatial profile are 0.5 cm and 1.0 cm, respectively. The width and length at the base of the spatial profile are 1.5 cm and 2.0 cm, respectively. The effective beam area is $a = 1.57$ cm^2 (the volume divided by the height). The spatial amplitude profile used in the calculations is a slice through the length of the spatial profile which is represented by a trapezoid of unity amplitude, a length at the top of 1 cm and a length at the base of 2 cm. The spatial amplitude profile of the pump used in the calculations is diagrammed in Fig. 3.7 as $S_P(y)$. The spatial amplitude profile is the signal that would be measured with a photodiode array positioned at the center of the pump laser beam. The normalized temporal profile for the pump laser is again the XeCl excimer laser pulse shown in Fig. 3.5a. The pump intensity propagates in the x direction across the width of the sample cell, which is shown as the amplifier cell in Fig. 3.7. The dye laser oscillator in Fig. 3.7 should be ignored for now. The position dimension of the spatial amplitude profile of the pump beam is taken to be along the length of the sample cell in the y direction, as shown in

Fig. 3.7. Thus the pump intensity in space and time is

$$I_P(x = 0, y, t) = \frac{E_P V_P(t) S_P(y)}{a \Delta t_P} \,, \qquad (3.26)$$

where $V_P(t)$ is the amplitude of pump laser temporal profile (Fig. 3.5a) at time t, $S_P(y)$ is the amplitude of the spatial profile (trapezoid) at position y along the length of the cell and

$$\Delta t_P = \int_0^\infty V_P(t) dt$$

as before. Because the steady state approximation in effect disconnects the temporal profile from the spatial dimensions, including the temporal profile in the numerical integration of (3.25) is quite simple. For a laser pulse with measured energy, E_P, the fluence is calculated at each point along the digitized spatial amplitude profile, $S_P(y)$. Each point along $S_P(y)$ has an intensity associated with each point across the digitized temporal profile, $V_P(t)$, computed with (3.26). During the integration of (3.25) across the width, w, of the cell, a position y is chosen from which the intensity associated with each individual temporal slice of $V_P(t)$ is propagated from $x = 0$ to $x = w$ according to (3.25). This is repeated for each value of y in $S_P(y)$. The transmitted intensity is finally obtained by summing all the individual contributions from $S_P(y)$ and normalizing the result to the area of $S_P(y)$.

The effect of the excited state absorption $S_2 \leftarrow S_1$ on the transmission of the pump laser beam through a dye sample can be demonstrated by integrating (3.25), with the sample conditions described above, for different values of R. Figure 3.10 shows plots of transmission vs. pump laser intensity for three values of R.

Transmission vs. intensity data calculated with the value of R listed in Table 3.2 is shown as boxes in Figs. 3.11a–c. Employing the temporal and spatial profiles and the two-parameter fitting routine (Sect. 3.4), a best fit to the data is obtained for $I_{PS} = 2.58$ MW/cm^2 and $R = 0.40$, which agrees exactly with the values in Table 3.2. Since the data and the fit are generated with the same differential equation, the fit is expected to be exact. If the temporal profile is used but instead of using the spatial profile $S_P(t)$ explicitly in the fitting routine, the beam area a alone is employed, then the best fit to the data (Fig. 3.11b) yields $I_{PS} = 3.21$ MW/cm^2 and $R = 0.41$. If neither the temporal nor spatial profiles are used explicitly, but only the FWHM approximation for pulse width and the beam area, a, are used in the fitting, then the best fit (Fig. 3.11c) yields $I_{PS} = 5.22$ MW/cm^2 and $R = 0.43$. In the latter case the pump pulse laser intensity is calculated with (3.24), which is the standard method. However, the results in Figs. 3.11a–c demonstrate that ignoring the spatial and temporal profiles of a laser pulse can lead to erroneous results in what appear to be good fits to data. Also from these results it is apparent that I_{PS} is more strongly dependent than R on the exacting treatment of I_P in fitting the data. The value of

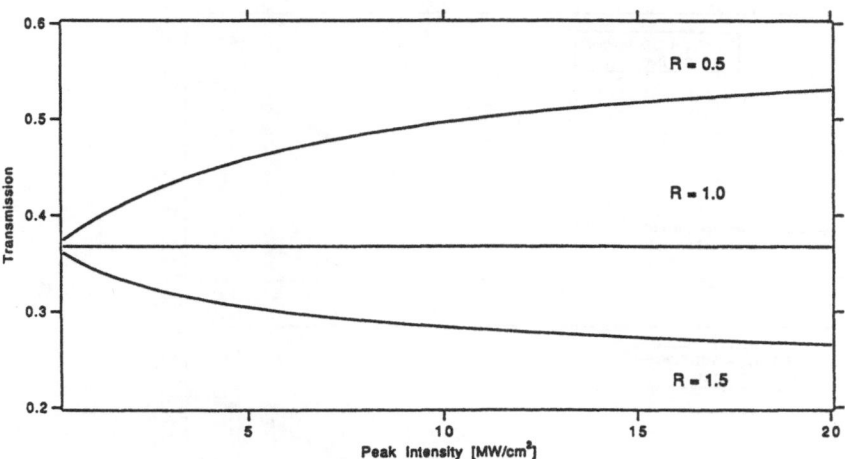

Fig. 3.10. Transmission vs. pump laser intensity through laser dye for three values of excited state absorption strengths

R determines the high intensity limit of the transmission and so is somewhat insulated from the specifics of I_P in the fitting routines.

The choice of solvent can also affect the performance of a dye laser without being optically active [3.98–100]. In the gain medium of a dye laser it is possible that the liquid or solid host, in which the laser dye is dissolved or suspended, is optically active and can affect the performance of the dye laser [3.74]. Although such an effect can usually be minimized by proper selection and purification of the host material, it is still feasible to analyze a dye sample with the effect present. For example, consider the case of a solvent which exhibits significant coherent two-photon absorption at the pump laser wavelength. This can be observed by measuring the transmission vs. pump laser intensity through the nascent solvent. The equation for propagation of the pump laser beam intensity through the solvent is

$$\frac{dI_P}{dx} = -n_S\sigma_S I_P - n_S\sigma_{2P} I_P^2 \ , \tag{3.27}$$

where n_S is the solvent density, σ_S is the solvent absorption cross section and σ_{2P} is the two-photon absorption cross section of the solvent. The solution of (3.27) gives an expression for transmission

$$T = \frac{e^{-n_S\sigma_S w}}{1 + (\sigma_{2P}/\sigma_S)\,I_P(1 - e^{-n_S\sigma_S w})} \ . \tag{3.28}$$

If in (3.27) the solvent absorption cross section is near zero, then the photon

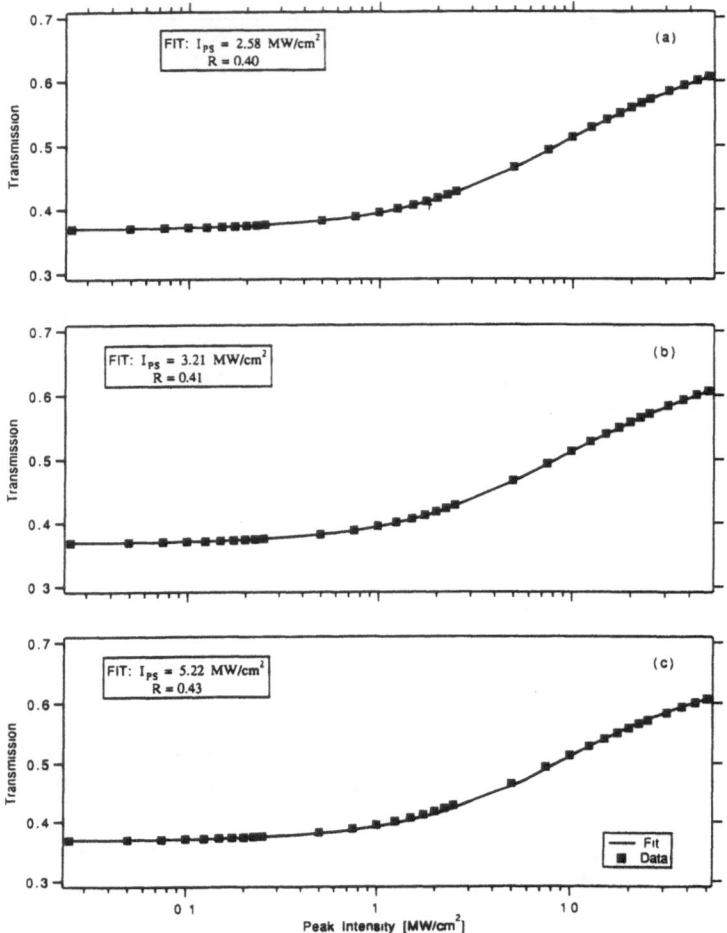

Fig. 3.11a–c. Semi-log plots of transmission vs. pump laser intensity through a laser dye for three different fitting approximations. Boxes are data and solid lines are theoretical fits

propagation equation is

$$\frac{dI_\mathrm{P}}{dx} = -n_\mathrm{S}\sigma_{2\mathrm{P}}I_\mathrm{P}^2 \tag{3.29}$$

and the transmission becomes,

$$T = \frac{1}{1 + n_\mathrm{S}\sigma_{2\mathrm{P}}wI_\mathrm{P}} . \tag{3.30}$$

A plot of (3.30) with $\sigma_{2\mathrm{P}} = 5 \times 10^{-48}$ cm^4 s and $n_\mathrm{S} = 7.06 \times 10^{21}$ cm^{-3} (the density of para-dioxane) is shown as boxes in Figs. 3.12a. This value for σ_{2P} is significantly higher than is found in common solvents and is chosen for illustrative comparisons. Cyclohexane exhibits a strong two-photon absorption which results in photochemical decomposition [3.101, 102]. The solid curve in Fig. 3.12a is a fit to the solvent transmission data with (3.30). This fit to the data is a simple procedure to determine the two-photon cross section of a sample.

Another process which occurs in solvents that almost exactly replicates the curve in Fig. 3.12a is laser beam scattering from suspended microscopic particles. Even the highest grade solvents are contaminated with dust or other particles. Distillation and filtration will remove these contaminants.

A simple test to determine if the solvent is a non-linear absorber (or scatterer) merely requires the measurement of the transmission of a solvent sample as function of pump laser intensity, as in Fig. 3.12a. If the measured transmission

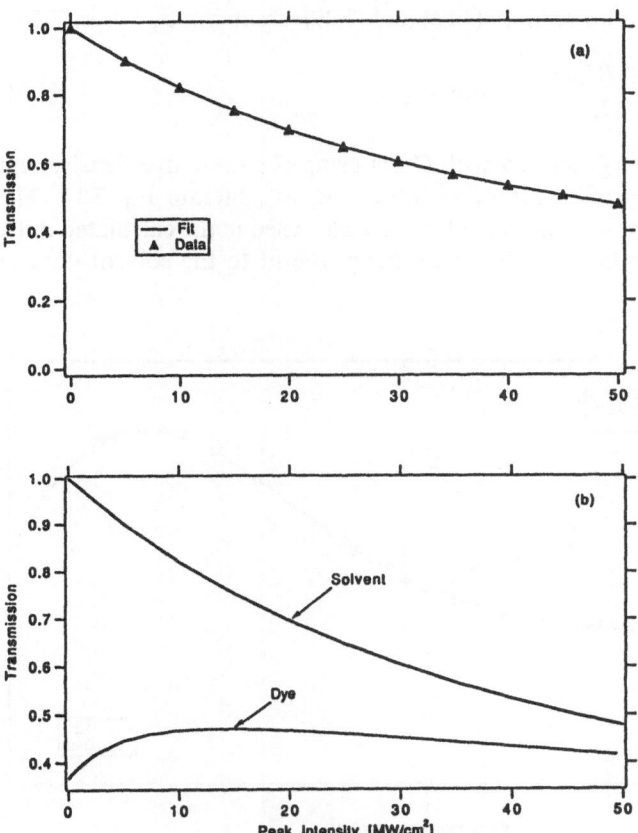

Fig. 3.12. Transmission vs. pump laser intensity; (a) pure non-linear absorbing solvent, (b) solvent and dye/solvent transmissions

68   C. Jensen

vs. intensity curve is flat over the intensity range of interest then the solvent in this regard is acceptable. Additional particles are often generated in the dye solution flow loops in dye lasers. Precautions must be taken to check for the effects of these particles and to filter them from the dye solution as described in Chap. 4.

If laser dye is dissolved in a non-linear absorbing solvent and the transmission vs. intensity is measured then the result is the lower curve in Fig. 3.12b. The transmission vs. pump intensity curve of the dye, instead of increasing with intensity as plotted in the upper curve of Fig. 3.10, turns over and decreases in value at the higher intensities in response to the optical effect of the solvent. A simple test for the optical quality of the dye solution in a dye laser is provided by the measurement of the transmission vs. pump laser intensity with a sample of the dye solution. If a curve with the shape found in Fig. 3.12 is measured then the dye solution is contaminated with particles, non-linear absorbers or both.

If the non-linear absorption loss in the solvent cannot be removed, the laser dye can still be accurately analyzed. The pump-laser photon propagation equation becomes a combination of (3.25) and (3.29) (assuming $\sigma_S = 0$ in (3.27))

$$\frac{dI_P}{dx} = -\frac{n\sigma_{01}I_P(1 + RI_P/I_{PS})}{1 + I_P/I_{PS}} - n_S\sigma_{2P}I_P^2 . \tag{3.31}$$

Transmission data were generated with (3.31) using the same dye density and spatial and temporal profiles for I_P as were used to generate Fig. 3.11. The solvent optical constants used in Fig. 3.12a were also used in the calculated data. The data are shown as boxes in Fig. 3.13. From the fit to the solvent data in

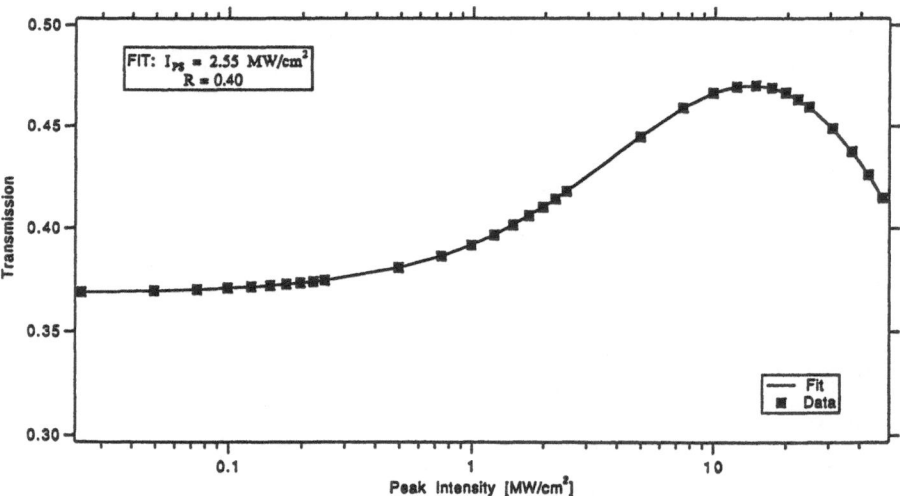

Fig. 3.13. Semi-log plot of transmission vs. pump laser intensity of dye dissolved in a non-linear absorbing solvent

Fig. 3.12a, the optical constants for the solvent are known. Using the fitting routine in Sect. 3.4 to fit the data in Fig. 3.13 to (3.31), the solid curve in Fig. 3.13 is obtained corresponding to $I_{PS} = 2.55$ MW/cm^2 and $R = 0.40$. The slight error in the fitted value for I_{PS} (compared to Table 3.2) results from using a high value for the error limit (Sect. 3.4) of the fit in order to reduce computation time. The versatility of numerically fitting directly to a differential equation allows known terms for parasitic processes to be added directly to the equation, which may render it analytically insoluble, without affecting the accuracy or simplicity of the fit to experimental data.

3.3.3 Gain

The last two optical constants for the laser dye to be found (Table 3.1) are the saturation intensity for stimulated emission, I_S, and the ratio of the excited state absorption cross section to the stimulated emission cross section, R_e. The data needed to determine their values result from the measurements of the amplification of a dye laser probe beam through a dye laser amplifier as a function of the probe beam incident intensity. A block diagram for the experiment is the same as described previously and is shown in Fig. 3.7. The dye laser amplifier is optically pumped along its length and through its width by the pump laser beam which is maintained at a fixed pulse energy. The probe laser beam from a dye laser oscillator propagates through the length of the amplifier. The probe laser pulse energies are measured before and after the amplifier. The temporal and spatial profiles of the probe beam are $V(t)$ and $S(x)$ respectively. The overlap of the probe and pump temporal profiles are shown in Fig. 3.5a. Simulated data is numerically generated with (3.20) which is coupled to (3.19). Given that the pump and probe beams propagate in orthogonal directions, the two equations that describe them can be solved only by numerical means.

Data generated with a peak pump intensity $I_P = 10$ MW/cm^2 and a peak probe intensity varying from $I = 0$ to 20 MW/cm^2 are plotted as boxes in Fig. 3.14. Fitting the data to (3.19, 20) by the method in Sect. 3.4, the solid curve in Fig. 3.14 was obtained. The fitted values are $I_S = 2.12$ MW/cm^2 and $R_e = 0.20$ as is expected from Table 3.1. As was shown in the analysis of transmission data (Sect. 3.3.2) the spatial and temporal profiles must be used explicitly in the numerical integrations to fit the data correctly. The use of FWHM estimations of the temporal pulse width and spatial beam areas to calculate laser pulse intensities leads to serious errors.

The determination of I_S and R_e demonstrates, in particular, the advantage of measuring the optical constants in the photon propagation equations directly by simple experiments. In the conventional analysis methods I_S is calculated from its component parts σ_e and τ with (3.16b). The independent measurement of σ_e is especially tedious since it requires the measurement of the quantum yield for fluorescence and the fluorescence lifetime and σ_e must be known before σ_{12}^L can be determined [3.54]. It is therefore a significant result that, with the saturation analysis method, R_e can be determined independently from I_S.

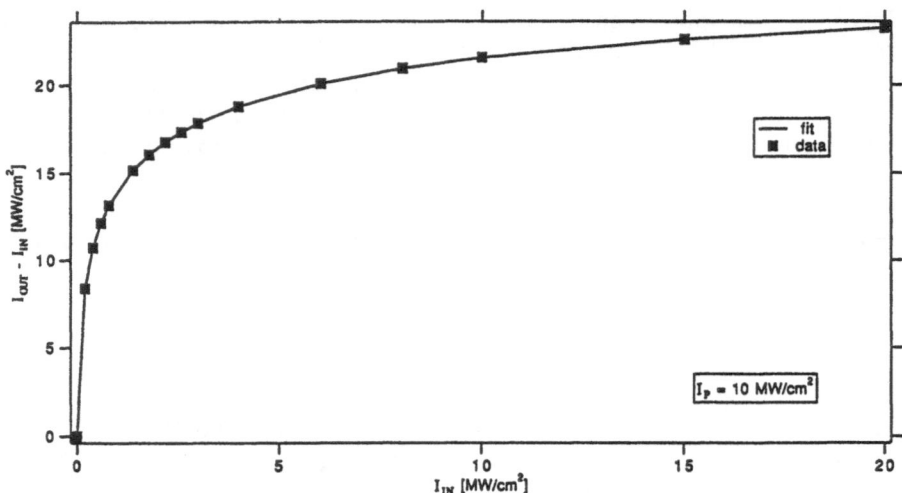

Fig. 3.14. Probe laser extraction intensity vs. probe laser input intensity for a dye laser amplifier

By completing the series of experiments listed in Table 3.1 (or at least numbers 1, 3 and 4) all of the optical constants for a laser dye/solvent combination, $\sigma_{01}, \sigma_{01}^{L}, I_{PS}, I_{S}, I_{LS}, R$ and R_{e}, can be measured accurately and with common laboratory equipment. The equations (3.19, 20) can now be used to predict and optimize the performance of a dye laser amplifier design employing the laser dye/solvent combination as a gain medium (Sect. 3.6).

The rest of the optical constants in Table 3.2 can be computed from the optical constants measured in the experimental series. From I_{PS} (Sects. 3.3.1 or 3.3.2) and (3.16a) the fluorescence lifetime $\tau = 1.00$ ns is obtained. The value of R (Sect. 3.3.2) and (3.21a) yields $\sigma_{12} = 1.00 \times 10^{-16}$ cm^2. The absorption cross section σ_{01}^{L} and τ plugged into (3.16c) (or using $I_{LS} = I_{PS} \times \sigma_{01}/\sigma_{01}^{L}$) gives the saturation intensity for probe beam absorption, $I_{LS} = 3310$ MW/cm^2. Finally, from I_{S} and R_{e} and (3.16b) and (3.21b), the stimulated emission cross section $\sigma_{e} = 2.34 \times 10^{-16}$ cm^2, and the excited state absorption cross section $\sigma_{12}^{L} = 0.47 \times 10^{-16}$ cm^2 at the dye laser wavelength. With the known values of the optical constants in Table 3.2 and the fluorescence band in Fig. 3.8, the quantum yield for fluorescence and the intersystem crossing rate, k_{ST}, can also be determined (Sect. 3.5).

With the data obtained so far, the wavelength dependence of σ_{e} can be calculated (Sect. 3.5), however the wavelength dependence of σ_{12}^{L}, if needed, must be measured by repeating the gain experiment at different probe laser wavelengths.

3.4 The Fit of Data to Differential Equations

The measurements of the saturation intensities and cross section ratios require fits of the experimental data to differential equations. The determination of I_{PS} and R requires a fit of the transmission data (Sect. 3.3.2) to (3.25). Similarly the values of I_S and R_e are deduced from a fit of the amplifier gain data (Sect. 3.3.3) to the coupled equations (3.19) and (3.20). The fit of experimental data to numerically integrated differential equations is quite simple in that it follows the same format as linear least squares fitting. Let DE be the differential equation to which the data is to be fit. Take f to be the integral of DE, presumably by a numerical integration technique,

$$f = \int_0^b DE \, d\zeta \tag{3.32}$$

where ζ is the propagation direction variable and b is the range of integration. For each experimental datum y_i there is a theoretical point f_i calculated from (3.32). It is assumed that the best estimate of the model parameters is obtained by minimizing the quantity

$$\chi^2 = \sum_{i=1}^N (y_i - f_i)^2 \; . \tag{3.33}$$

It is also assumed that all the experiments have the same standard deviation [3.103]. The two adjustable parameters in (3.25) and (3.31) for fitting transmission data are R and I_{PS}. The two adjustable parameters in (3.19) and (3.20) for fitting gain data are R_e and I_S. For the following general treatment the adjustable parameters are called α and β. The first step is to linearize f_i by expanding it in a first order Taylor series in two variables, α and β

$$f_i = f_{0i} + \frac{\partial f_{0i}}{\partial \alpha} \Delta\alpha + \frac{\partial f_{0i}}{\partial I_{PS}} \Delta\beta \; , \tag{3.34}$$

where f_{0i} is the current guess at the value of f_i, $\Delta\alpha$ is the correction to the current value of the parameter α and $\Delta\beta$ is the correction to β. To minimize (3.33) with respect to these parameter corrections requires the solution of two equations

$$\frac{\partial \chi^2}{\partial \Delta\alpha} = 0 \tag{3.35}$$

and

$$\frac{\partial \chi^2}{\partial \Delta\beta} = 0 \; . \tag{3.36}$$

Substituting (3.34) into (3.33) and substituting the result into (3.35) and (3.36) gives two equations in the two unknowns $\Delta\alpha$ and $\Delta\beta$

$$\sum_{i=1}^{N}(y_i - f_{0i})\frac{\partial f_{0i}}{\partial\alpha} = \sum_{i=1}^{N}\left(\frac{\partial f_{0i}}{\partial\alpha}\right)^2\Delta\alpha + \sum_{i=1}^{N}\left(\frac{\partial f_{0i}}{\partial\beta}\right)\left(\frac{\partial f_{0i}}{\partial\alpha}\right)\Delta\beta \tag{3.37}$$

and

$$\sum_{i=1}^{N}(y_i - f_{0i})\frac{\partial f_{0i}}{\partial\beta} = \sum_{i=1}^{N}\left(\frac{\partial f_{0i}}{\partial\alpha}\right)\left(\frac{\partial f_{0i}}{\partial\beta}\right)\Delta\alpha + \sum_{i=1}^{N}\left(\frac{\partial f_{0i}}{\partial\beta}\right)^2\Delta\beta \ . \tag{3.38}$$

For convenience let

$$X_1 = \sum_{i=1}^{N}(y_i - f_{0i})\frac{\partial f_{0i}}{\partial\alpha}, \quad X_2 = \sum_{i=1}^{N}(y_i - f_{0i})\frac{\partial f_{0i}}{\partial\beta},$$

$$C_{11} = \sum_{i=1}^{N}\left(\frac{\partial f_{0i}}{\partial\alpha}\right)^2, \quad C_{12} = \sum_{i=1}^{N}\left(\frac{\partial f_{0i}}{\partial\beta}\right)\left(\frac{\partial f_{0i}}{\partial\alpha}\right),$$

$$C_{21} = \sum_{i=1}^{N}\left(\frac{\partial f_{0i}}{\partial\alpha}\right)\left(\frac{\partial f_{0i}}{\partial\beta}\right) \quad\text{and}\quad C_{22} = \sum_{i=1}^{N}\left(\frac{\partial f_{0i}}{\partial\beta}\right)^2 .$$

Equations (3.37) and (3.38) then become

$$X_1 = C_{11}\Delta\alpha + C_{12}\Delta\beta \tag{3.39}$$

and

$$X_2 = C_{21}\Delta\alpha + C_{22}\Delta\beta \ . \tag{3.40}$$

Solving the system of equations (3.39) and (3.40) for $\Delta\alpha$ and $\Delta\beta$ yields

$$\Delta\alpha = \frac{\begin{vmatrix} X_1 & C_{12} \\ X_2 & C_{22} \end{vmatrix}}{\begin{vmatrix} C_{11} & C_{12} \\ C_{21} & C_{22} \end{vmatrix}}, \quad \Delta\beta = \frac{\begin{vmatrix} C_{11} & X_1 \\ C_{21} & X_2 \end{vmatrix}}{\begin{vmatrix} C_{11} & C_{12} \\ C_{21} & C_{22} \end{vmatrix}} . \tag{3.41}$$

The values of α and β are corrected by

$$\alpha \text{ (corrected)} = \alpha + \Delta\alpha$$

and

$$\beta \text{ (corrected)} = \beta + \Delta\beta \ ,$$

and the corrected values for α and β are substituted into (3.34) and new values for $\Delta\alpha$ and $\Delta\beta$ are calculated. The process is repeated until

$$\frac{\Delta\alpha}{\alpha} \leqslant Er$$

and

$$\frac{\varDelta\beta}{\beta} \leqslant \text{Er} ,$$

where Er is the desired error limit. The fitting calculations in Sect. 3.3 used Er = 0.1%. The statistical measure of goodness-of-fit is estimated by the standard deviation in the usual way [3.103].

For the fit of pump laser beam transmission vs. I_P data in Sect. 3.3.2, $\alpha = R$, $\beta = I_{PS}$. For the predicted values of the data calculated with (3.32), $DE = (3.25)$ or (3.31), $d\zeta = dx$, and $b = w$. The partial derivative terms in (3.34) become

$$\frac{\partial f}{\partial R} = -\int_0^w \frac{n\sigma_{01} I_P^2}{I_{PS}(1 + I_P/I_{PS})} dx \tag{3.42}$$

and

$$\frac{\partial f}{\partial I_{PS}} = -\int_0^w \frac{n(1 - R)\sigma_{01} I_P^2}{I_{PS}^2(1 + I_P/I_{PS})^2} dx , \tag{3.43}$$

where f becomes f_{0i} when an initial guess for R and I_{PS} is made. The initial guesses for R and I_{PS} in Sect. 3.3.2 were deliberately set 50–200% from the known values. The fitting routine usually ran through four or five iterations before the error limit (Er) was achieved. The numerical integration of (3.32), (3.42) and (3.43) can be performed in the same integration subroutine of the computer program.

The fitting of the gain data in Sect. 3.3 is slightly more complicated than that of transmission because the integration must be over two directions and two coupled differential equations, (3.19) and (3.20). The simplest method is to make I and I_P two-dimensional arrays in x and y and integrate (3.19) and (3.20) over a two-dimensional grid spanning the x–y plane of the amplifier shown in Fig. 3.7. The spatial amplitude profiles of the pump and probe beams are $S_P(y)$ and $S(x)$, respectively. It is assumed that $S_P(y)$ and $S(x)$ are coplanar x–y planes of symmetry which slice through the actual three-dimensional spatial profiles of the laser beams and accurately represent the spatial profiles in the numerical integrations. These are the normalized (to a peak value of one) amplitude profiles that would be measured by a photodiode array located in the center of a symmetric laser beam. The temporal profiles of the pump and probe laser pulses are $V_P(t)$ and $V(t)$, respectively. The overlap of the temporal profiles in time is shown in Fig. 3.5a. The profiles $S_P(y)$ and $S(x)$ are the initial conditions (at $t = 0$) for the numerical integration of the photon flux (intensity) equations (3.19) and (3.20). For each laser beam, the pulse energy combined with the spatial profile provides the laser fluence as a function of position in the amplifier. The temporal profile combined with the fluence determines the intensity as a function of time.

In the calculation of f with (3.32) DE becomes (3.20), $\zeta = y$, and $b = L$, the length of the amplifier. In (3.34) $\alpha = R_e$, $\beta = I_S$ and the partial derivatives of f are

$$\frac{\partial f}{\partial R_e} = -\int_0^L \frac{nI(\sigma_{01}I_P + \sigma_{01}^L I)}{I_S\left(1 + \frac{I_P}{I_{PS}} + \frac{I}{I_S} + \frac{I}{I_{LS}}\right)} dy \qquad (3.44)$$

and

$$\frac{\partial f}{\partial I_S} = -\int_0^L \frac{n(1 - R_e)(\sigma_{01}I_P I + \sigma_{01}^L I^2)\left(1 + \frac{I_P}{I_{PS}} + \frac{I}{I_{LS}}\right)}{I_S^2\left(1 + \frac{I_P}{I_{PS}} + \frac{I}{I_S} + \frac{I}{I_{LS}}\right)^2} dy \ . \qquad (3.45)$$

Because of the presence of both I_P and I in (3.19), (3.20), (3.44) and (3.45) these equations must be evaluated over the entire two-dimensional integration grid. The values of the probe laser output intensity of the amplifier, and the partial derivative terms are calculated by summing the contributions to the intensity from each slice of the spatial profile $S(x)$ at $y = L$ (after summing the contributions to each slice of $S(x)$ from each increment of $V(t)$). The result is normalized to the areas of spatial amplitude profiles and temporal profiles.

3.5 Quantum Yield

Using the experimental series of Table 3.1, all the optical parameters of the laser dye in Table 3.2 have been determined except for the intersystem crossing rate k_{ST}. Although k_{ST} is not usually germane to excimer laser-pumped dye lasers, it is to long pulse dye lasers described in Chap. 6, and to continuous wave lasers. To arrive at a value for k_{ST} another parameter of the laser dye must be determined, called the quantum yield of fluorescence. A quote from *Becker* [3.6] applies to laser dyes and states: "The quantum yield is the most difficult characteristic to measure."

The quantum yield, Q, can be defined in several equivalent ways but is simply stated as

$$Q = \frac{\text{number of photons emitted}}{\text{number of photons absorbed}} \ .$$

A direct measure of the quantum yield requires the careful measurement of the number of photons absorbed and emitted by a sample. Many approaches to such measurements have been developed [3.7, 104, 105] and as with fluorescence lifetime measurements, quantum yield measurements are encumbered by self absorption [3.106]. The quantum yield is also the probability that a photon is emitted from S_1 by the transition $S_0 \leftarrow S_1$, which can define a probability function, $\gamma(\lambda)$, by

$$Q = \int_0^\infty \gamma(\lambda) d\lambda \ . \tag{3.46}$$

The probability function $\gamma(\lambda)$, can be related to the fluorescence band $f(\lambda)$ in Fig. 3.8. Normalizing $f(\lambda)$ so that the wavelength function $E(\lambda)$ is given by

$$E(\lambda) = \frac{f(\lambda)}{\int_0^\infty f(\lambda) d\lambda} \ , \tag{3.47}$$

then

$$\int_0^\infty E(\lambda) d\lambda = 1 \tag{3.48}$$

so $\gamma(\lambda) = Q \times E(\lambda)$. The wavelength function, $E(\lambda)$, is easily obtained from the fluorescence band since the amplitude of the fluorescence band does not have to be in absolute units. Figure 3.15a shows the $E(\lambda)$ corresponding to the $f(\lambda)$ in Fig. 3.8.

   Peterson [3.41, 42, 44, 45, 107] has derived the wavelength dependence of the stimulated emission cross section, σ_e, in terms of the quantum yield and $E(\lambda)$

$$\sigma_e(\lambda) = \frac{QE(\lambda)\lambda^4}{8\pi\eta^2 C\tau} \ , \tag{3.49}$$

where η is the index of refraction of the solvent and c is the velocity of light. It is apparent from (3.49) why the conventional method of determining σ_e requires independent measurements of the quantum yield, the fluorescence lifetime and the fluorescence band. However, if (3.49) is rearranged and using (3.16b), then

$$Q = \frac{8\pi\eta^2 C}{E(\lambda)\lambda^4 I_S} \ . \tag{3.50}$$

With $E(\lambda)$ computed from the fluorescence band and I_S measured at a single wavelength ($\lambda = 400$ nm) in the gain experiment of Sect. 3.3.3, the quantum yield can be calculated directly from (3.50). At $\lambda = 400$ nm in Fig. 3.15a, $E(400 \text{ nm}) = 0.0164 \text{ nm}^{-1}$ ($1.64 \times 10^5 \text{ cm}^{-1}$), $\eta = 1.42$ (para-dioxane), $c = 3 \times 10^{10}$ cm/s and $I_S = 2.12 \text{ MW/cm}^2$ (4.27×10^{24} photons/cm^{2}s). Plugging these values into (3.50) gives $Q = 85\%$.

   With the value of Q now known and the value of τ determined from Sects. 3.3.1 or 3.3.2, then (3.49) can be used to calculate the value σ_e for all λ in the fluorescence band as shown in Fig. 3.15b.

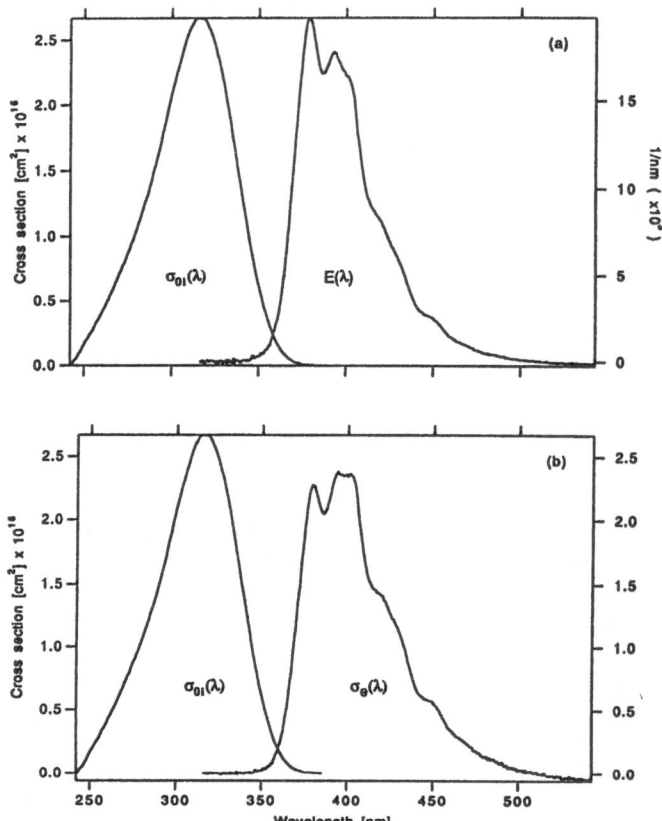

Fig. 3.15. Absorption and emission bands: (a) absorption band (left) and normalized fluorescence band, left ordinate corresponds to absorption band, (b) absorption (left) and stimulated emission bands

The quantum yield can also be expressed as the ratio of the rate at which S_1 radiates to the sum of all rates which depopulate S_1

$$Q = \frac{k_{\text{rad}}}{k_{\text{rad}} + k_{\text{ST}} + k_{\text{IC}} + \ldots} \ . \tag{3.51}$$

Noting that $k_{\text{rad}} = 1/\tau_{\text{rad}}$ and assuming that the intersystem crossing rate k_{ST} is the only significant non-radiative process, then (3.51) becomes

$$Q = \frac{1/\tau_{\text{rad}}}{1/\tau_{\text{rad}} + k_{\text{ST}}} \ . \tag{3.52}$$

Substituting (3.8) into (3.52)

$$Q = \frac{\tau}{\tau_{rad}} \cdot \qquad (3.53)$$

Rearranging (3.52)

$$Q = \frac{1}{1 + k_{ST}\tau_{rad}} , \qquad (3.54)$$

from which is obtained

$$k_{ST} = \frac{1}{\tau_{rad}} \left(\frac{1}{Q} - 1 \right) \qquad (3.55a)$$

or

$$k_{ST} = \frac{1}{\tau}(1 - Q) \cdot \qquad (3.55b)$$

Using the known values for τ and Q in (3.53), $\tau_{rad} = 1.18$ ns. From (3.55b) $k_{ST} = 1.50 \times 10^8$ s^{-1} which completes the recovery of the entries in Table 3.2. If other non-radiative depopulation processes occur from S_1 then (3.55b) becomes an upper bound for the intersystem crossing rate

$$k_{ST} \leqslant \frac{1}{\tau}(1 - Q) \cdot \qquad (3.56)$$

In general, this approximation for estimating k_{ST} is most accurate for short wavelength-emitting dyes but becomes a poor approximation for infrared-emitting dyes. In the long wavelength regime, where the excited and ground states are energetically close together, radiationless transitions from the excited electronic state to excited vibrational states of the ground electronic state (internal conversion) will dominate the non-radiative quenching of the excited electronic level. For these long wavelength dyes, additional measurements must be made to determine the intersystem crossing rate constant k_{ST}.

3.6 Amplifier Design

In this section a simple model is described for estimating the design parameters of a double-sided laser-pumped dye laser amplifier. The model uses some simplifying assumptions which make it easy to use but the results should be considered approximate. It is assumed that the laser dye and solvent combination has been analyzed (Sect. 3.3) and the optical constants are known. The design of an amplifier can be improved by calculating the performance of the amplifier with the photon flux equations (3.19, 20). The amplifier design model parameters can be adjusted to be in closer agreement with the photon flux calculations and can then be used for optimizing the design of an amplifier for a set of specifications and constraints.

The amplifier is rectangular and transversely pumped along the side(s) with an excimer or other pulsed laser source and the probe beam, from a dye laser oscillator or previous amplifier, enters one end and exits the opposite end as diagramed in Fig. 3.7. From the experimental measurement series and analysis of Sects. 3.3.1–3, all the relevant optical constants of the dye laser gain medium listed in Table 3.2 are known. The problem in amplifier design is how to trade off the physical dimensions of the amplifier cell with the dye concentration, beam quality and laser intensities involved to provide the desired amplification of the input laser beam without damaging the amplifier optics.

The performance of an amplifier of given physical dimensions is calculated with (3.19, 20). Consider an amplifier with a 4 mm width and height and a length of 2 cm. The pump and probe beams are assumed to be symmetric about the center plane of the amplifier, both with spatial profiles represented by a square wave so that the effects of the spatial profile do not mask the following considerations.

The variables in (3.19, 20) are the dye density, n, and the pump and probe intensities, I_P and I. When the amplifier is pumped on one side with $I_P = 10 \, \text{MW/cm}^2$ the output intensity of the probe beam, I_{out}, can be calculated for a range of probe input intensities, I_{in}. The dependence of I_{out} on the dye density is shown in Fig. 3.16a. Clearly the output intensity of the probe beam scales in an increasing manner with the dye density. It is assumed, for illustrative purposes, that the range of densities shown in Fig. 3.16a are within the low density range of the laser dye. At higher concentrations the effects of concentration quenching and self absorption can also derogate from the performance of the dye laser amplifier. Since these effects are not included in the model they must be checked experimentally. The requirement of output beam uniformity can place the greatest constraint on dye concentration.

Similarly the pump intensity can be varied to affect the ouput intensity of the amplifier. With the dye density fixed at $n = 2.0 \times 10^{16} \, \text{cm}^{-3}$ (3.3×10^{-5} M) and the pump intensity varied in units of the pump saturation intensity, the results in Fig. 3.16b are obtained. For high amplification one wants to operate the amplifier with as high a dye density and pump intensity as is feasible. The limits on pump intensity are imposed by considerations of beam uniformity, excited state absorption, efficiency and optical damage.

It is interesting to note the effects of excited state absorption in the dye on the amplifier performance. The relative effects of excited state absorption by the pump and probe beams can be seen in Fig. 3.17. The top triplet of curves is obtained with $n = 2 \times 10^{16} \, \text{cm}^{-3}$, $I_P = 10 \, \text{MW/cm}^2$ and $R_e = 0$. The bottom family of curves result from $R_e = 0.20$. The three curves in each family correspond to three different values of R. As expected, the excited state absorption at the probe laser wavelength has a much more dramatic effect on the output

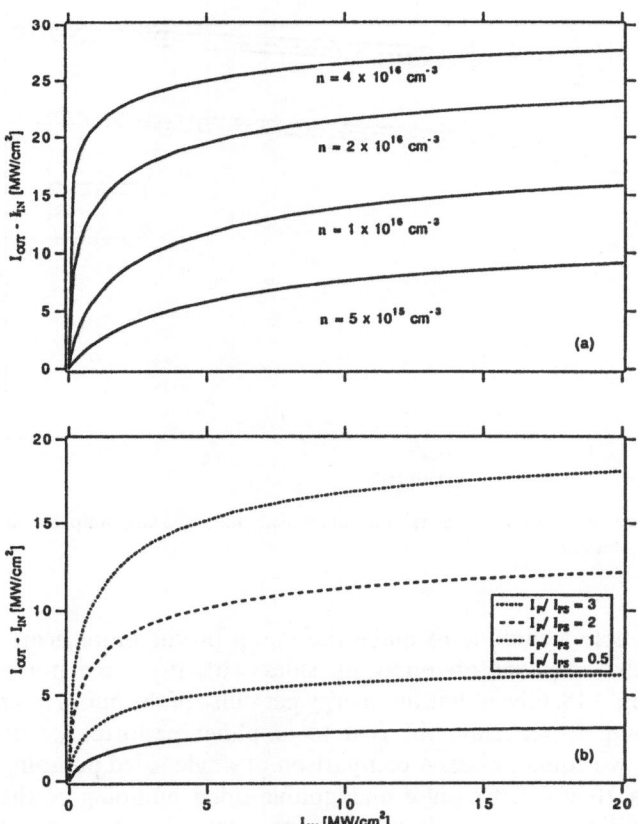

Fig. 3.16. Probe laser extraction intensity vs. probe laser input intensity for a dye laser amplifier with (**a**) different dye concentrations and (**b**) different pump laser intensities in values of the pump saturation intensity

intensity than the excited state absorption at the pump laser wavelength. The excited state absorption of either beam excites the molecule to S_2 where it thermally relaxes to S_1 and again becomes available for stimulated emission. The difference between the two processes is that an excited state absorption at the dye laser wavelength costs a photon directly from the output intensity, whereas the effect of excited state absorption at the pump laser wavelength is proportional to the efficiency of the laser. Here, the efficiency is taken to be the extraction power ($P_{out}-P_{in}$) relative to the incident power of the pump beam.

It is usually desirable that the spatial profile of the amplified probe beam exciting the amplifier should closely resemble that of the input. The high degree of amplification afforded by operating the amplifier with a high dye density and a high pump intensity will cause a severe distortion of the spatial profile. This results from the fact that the pump intensity is not uniform across the width of

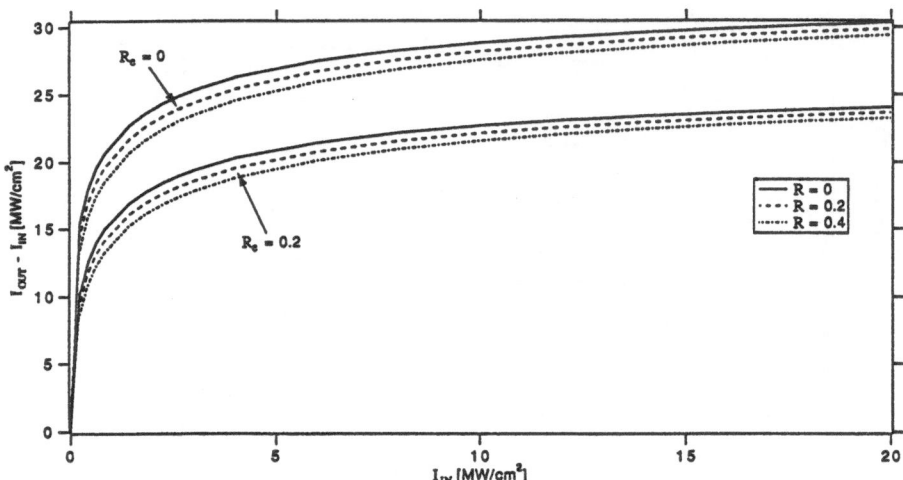

Fig. 3.17. Probe laser extraction intensity vs. probe laser input intensity for a dye laser amplifier for two values of R_e and three values of R

the amplifier cell. A simple technique to make the pump profile more nearly uniform is to pump the amplifier on opposing sides with the same pump intensity as shown in Fig. 3.18. Given that the energy per pulse of the pump laser is limited, it is important to ascertain the cost to amplifier performance by splitting the pulse into two equal pulses. A comparison of single-sided pumping of the amplifier with a 10 MW/cm² pulse and double-sided pumping of the amplifier with two 5 MW/cm² pulses (all with the same spatial and temporal profiles) is shown in Fig. 3.19. When the intensity of the pump beams exceeds the absorption saturation intensity, the cost of employing double-sided pumping is

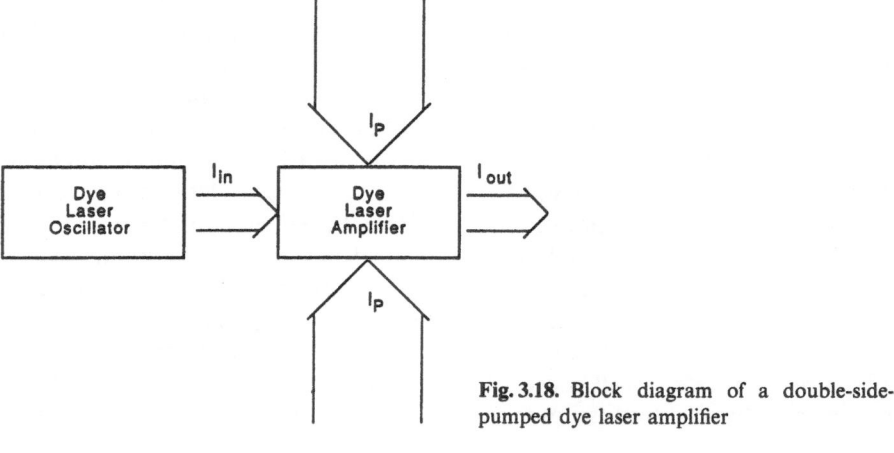

Fig. 3.18. Block diagram of a double-side-pumped dye laser amplifier

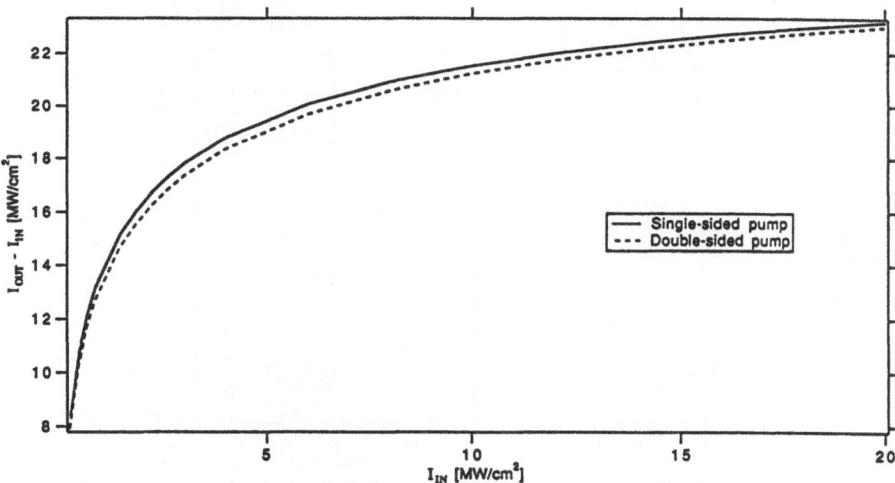

Fig. 3.19. Probe laser extraction intensity vs. probe laser input intensity for a dye laser amplifier excited by a single-side pump (10 MW/cm²) and a double-side pump (5 MW/cm² each)

small compared to the benefit gained from improved beam quality (at low dye density).

The effect of dye density on the beam quality can be seen from plotting the profiles of the pump and probe beams at the two ends of the amplifier. Figure 3.20 shows plots of pump and probe profiles for the amplifier under the conditions $I_P = 5$ MW/cm² (each side) and $I_{in} = 0.5$ MW/cm² for four different dye densities, calculated with (3.19, 20). In each graph in Fig. 3.20 the ordinate is the peak intensity for both the pump and probe pulses and the abscissa is the position along the width of the amplifier. I_{out} labels the probe intensity profile at the output end of the amplifier, $I_{P\,in}$ is the pump beam profile across the input end and $I_{P\,out}$ is the pump beam profile across the output end of the amplifier. The difference between $I_{P\,in}$ and $I_{P\,out}$ is the intensity lost from the pump beam to provide excitation of the dye molecules in the amplifier. As the dye density increases the pump intensity exciting the sides of the amplifier decreases, signifying more absorption by the dye solution. Also, as the dye density increases the probe output intensity increases but the uniformity of the probe beam spatial profile is lost. If the dye cell, with the conditions of Fig. 3.20d, was placed inside an oscillator optical cavity the output spatial profile would probably show two narrow laser beams propagating along the side walls of the cell and no intensity in the center. This case approaches the worse conditions for double-side-pumped dye lasers. The appropriate case will be a compromise between the desired amplification and the limitations represented by Fig. 3.20. The important effects of diffraction by the walls at the ends of the amplifier have been ignored for simplicity.

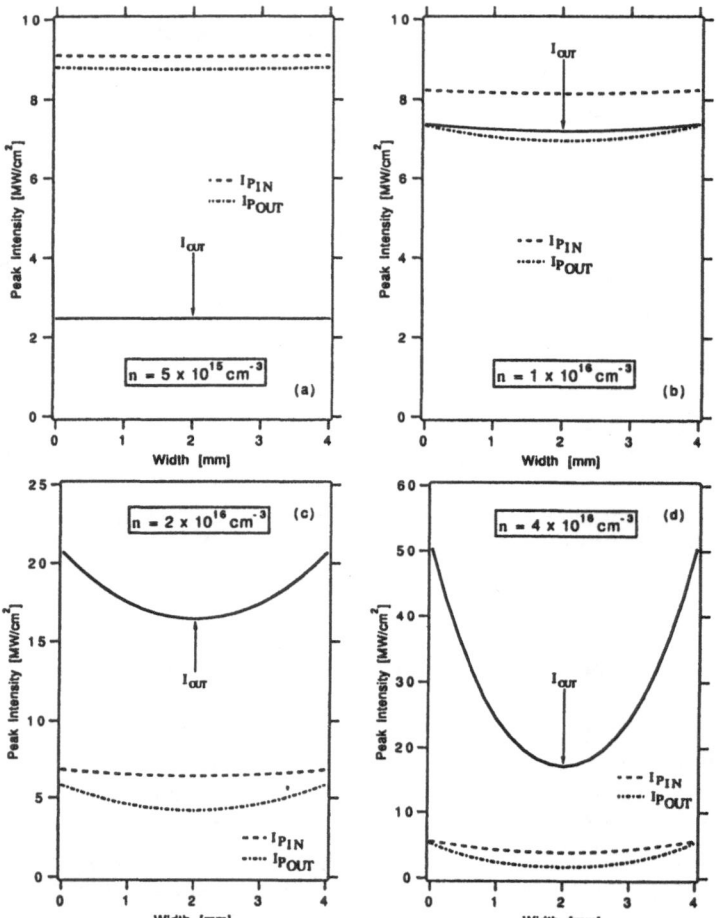

Fig. 3.20. Intensity profiles of pump and probe laser beams across input and output ends of a dye laser amplifier for four dye concentrations

The non-uniformity of the probe beam output spatial profile (resulting from a uniform input profile) can be exploited to improve the uniformity of an input beam which has a centrally peaked spatial profile. If the intensity of the input probe laser beam at the sides of the amplifier is lower than the intensity at the center, the non-uniform pump laser intensity profile across the width of the amplifier can cause the wings of the probe input profile to amplify more than the center region. This effect occurs anyway if the intensity at the wings of the probe input profile are in the small signal gain regime and the intensity in the central region of the profile is in the saturated amplification regime.

3.6.1 Double-Sided Pumped Amplifier

The double-sided pumped dye laser amplifier offers some obvious advantages over a single-side pumped amplifier. The physical dimensions of such an amplifier depend on the available probe beam input pulse energy, the desired output pulse energy, the available pump beam pulse energy and the damage threshold intensity of the amplifier cell optics. When an amplifier with near uniform pumping across its width is considered, then a simple gain theory can provide a rough estimate for physical dimensions.

In the derivation of the gain theory, the case where there is no excited state absorption at either the pump or probe laser wavelengths will initially be considered. The gain theory will be incorporated into a simple model for estimating the physical dimensions of an amplifier and the results will be compared to a more exacting calculation from (3.19, 20). In this simplified model the equations for propagation of the probe beam intensity and excited state density become

$$\frac{dI}{dy} = n_1 \sigma_e I \ , \tag{3.57}$$

$$\frac{dn_1}{dt} = n_0 \sigma_{01} I_P - \frac{n_1}{\tau} - n_1 \sigma_e I \ , \tag{3.58}$$

and

$$n = n_0 + n_1 \ . \tag{3.59}$$

Before solving this system of equations the relationship of gain in the amplifier to (3.57) can be shown. Separating variables in (3.57)

$$\int_{I_{\text{in}}}^{I_{\text{out}}} \frac{dI}{I} = \int_0^L n_1 \sigma_e dy \ ,$$

and integrating over the length L of the amplifier, assuming uniform pumping along the width and length,

$$\ln \frac{I_{\text{out}}}{I_{\text{in}}} = n_1 \sigma_e L = G \tag{3.60}$$

where G is the gain of the amplifier. If the stage gain of the amplifier is defined as

$$G_{\text{stage}} \equiv \frac{I_{\text{out}}}{I_{\text{in}}} \ , \tag{3.61}$$

then

$$G = \ln G_{\text{stage}} \ . \tag{3.62}$$

Returning to the system of equations (3.55–57) and assuming the steady state

$(dn_1/dt = 0)$, the solution of (3.56, 57) gives

$$n_1 = \frac{nI_P/I_{PS}}{1 + I_P/I_{PS} + I/I_S} .$$ (3.63)

Substituting (3.63) into (3.57) produces the intermediate result

$$I_S\left(1 + \frac{I_P}{I_{PS}}\right)\ln \frac{I_{out}}{I_{in}} + (I_{out} - I_{in}) = n\sigma_{01}LI_P .$$ (3.64)

Substituting (3.63) in (3.60) gives the expression for G

$$G = \frac{n\sigma_e LI_P/I_{PS}}{1 + I_P/I_{PS} + I/I_S} .$$ (3.65)

The maximum gain available in the gain medium would be evident when the upper state population was equal to the total density ($n_1 = n$). The maximum gain is defined from (3.60) as

$$G_M \equiv n\sigma_e L .$$ (3.66)

Substituting (3.66) into (3.65) gives the gain in terms of the maximum gain

$$G = \frac{G_M I_P/I_{PS}}{1 + I_P/I_{PS} + I/I_S} .$$ (3.67)

The small signal gain of the amplifier is the gain (3.67) when $I \to 0$, and is given by

$$G_0 = \frac{G_M I_P/I_{PS}}{1 + I_P/I_{PS}} .$$ (3.68)

Solving (3.68) for I_P and substituting the result into (3.64) gives the expression for the small signal gain in terms of the probe intensities

$$\ln \frac{\phi_{out}}{\phi_{in}} + \left(1 - \frac{G_0}{G_M}\right)(\phi_{out} - \phi_{in}) = G_0 ,$$ (3.69)

where

$$\phi_{out} = \frac{I_{out}}{I_S}, \phi_{in} = \frac{I_{in}}{I_S} \quad \text{and} \quad \phi_P = \frac{I_P}{I_{PS}}$$ (3.70)

are the intensities normalized to the saturation intensities. From (3.68) it is seen that the ratio G_0/G_M depends only on the pump intensity, hence this ratio can be called R_G

$$R_G = \frac{I_P/I_{PS}}{1 + I_P/I_{PS}} ,$$ (3.71)

so that (3.69) becomes

$$\ln \frac{\phi_{\text{out}}}{\phi_{\text{in}}} + (1 - R_G)(\phi_{\text{out}} - \phi_{\text{in}}) = G_0 \ . \tag{3.72}$$

The reason for developing the theory in terms of the small signal gain is that the small signal gain is easily related to design parameters, whereas the gain is not.

The design of an amplifier with a specified stage gain is now considered. As an example, the desired design is taken to be a rectangular amplifier which employs a dye with the parameters in Table 3.2 except, for now, the excited state cross sections σ_{12} and σ_{12}^{L} are equal to zero. A dye laser oscillator provides the probe laser pulses to be amplified by traversing the length of the amplifier. The pulse energy of the input probe laser is chosen to be 1 mJ, which is a common output energy from dye laser oscillators. The desired output energy of the amplifier is arbitrarily chosen to be 40 mJ, so therefore the stage gain is 40. It is also assumed that the peak output intensity from the amplifier should be limited to 20 MW/cm^2 to prevent optical damage.

The optical damage threshold of the wetted windows of the amplifier depends on the optically induced decomposition properties of the solvent and dye. Since these materials are predominantly organic they decompose relatively easily leaving carbon deposits and other insoluble organic deposits on the windows. Additional contaminants which attach to the windows are particles generated by the circulation pumps. These window deposits absorb laser radiation and substantially lower the optical damage threshold intensity of the optical surfaces from the value measured in air. Some hydrocarbon solvents such as cyclohexane cause severe problems of optical damage whereas alcohols, ethers and water are much less deleterious.

The peak intensity of the probe pulse is calculated from

$$I = \frac{E_{\text{L}}}{\Delta t_{\text{L}} A_{\text{end}}} \tag{3.73}$$

where E_{L} is the probe laser pulse energy, A_{end} is the spatial area of the probe beam which is taken to be equal to the area of each end of the amplifier, and Δt_{L} is the temporal pulsewidth of the probe pulse and is equal to the integral of the normalized temporal profile in Figs. 3.5a, and 3.7, i.e.

$$\int_0^\infty V(t)dt \cong 16 \text{ ns} \ .$$

Rearranging (3.73), the area of the end of the amplifier is

$$A_{\text{end}} = \frac{40 \text{ mJ}}{16 \text{ ns}} \frac{1}{20 \text{ MW/cm}^2} = 0.125 \text{ cm}^2 \ .$$

Again from (3.73), the input peak intensity is calculated to be $I_{in} = 0.5 \text{ MW/cm}^2$. Assuming the end area is square then the width and height are $w = h = 0.34$ cm.

To estimate the density of the dye molecules in the dye solution, a rule-of-thumb is invoked which states that the Beers law transmission, at the pump laser wavelength, through the width of the dye solution in the amplifier should be between 10% and 20%. It is assumed here that the dye density should be low enough to maintain a nearly uniform spatial profile for the probe beam. The estimation of the best transmission value depends on the saturation and excited state absorption properties of the dye. With a value of 15% chosen, then Beers law

$$\ln T = - n\sigma_{01} w$$

is solved for n to get $n = 2.4 \times 10^{16} \text{ cm}^{-3}$.

For independent reasons, such as optical damage threshold or dye photo-chemical stability, the pump laser peak intensity is required to be only 5 MW/cm^2 incident on each side of the amplifier. From (3.71), $R_G = 0.659$. Substituting the above values into (3.72), $G_0 = 6.88$ and subsequently $G_M = G_0/R_G = 10.4$. Solving (3.66) for the length of the amplifier

$$L = \frac{G_M}{n\sigma_e} = 1.86 \text{ cm} \ .$$

With the dimensions of the amplifier cell known, the area of the side can be calculated, $A_{side} = 0.658 \text{ cm}^2$. The required pump pulse energy incident on each side of the amplifier is

$$E_P = I_P \Delta t_P A_{side} = 53 \text{ mJ} \ .$$

To check the performance of the amplifier with greater accuracy, (3.19, 20) are input with the current parameter values and used to calculate the amplifier output intensity over a range of input intensities which are shown in Fig. 3.21a. For the input intensity calculated above, $I_{in} = 0.5 \text{ MW/cm}^2$, the output intensity calculated from (3.19, 20) is $I_{out} = 20.3 \text{ MW/cm}^2$ which is in reasonable agreement with the desired result (the temporal profiles were omitted). This result attests to a fortuitous choice in the rule-of-thumb estimation of the dye density. A different choice for the pump laser transmission, T, through the amplifier would result in a much worse disagreement between I_{out} values. For this reason the design parameters estimated with this simple model should always be checked with (3.19, 20). The corresponding spatial profiles for the pump and probe lasers at the ends of the amplifier are shown in Fig. 3.21b. The spatial uniformity of I_{out} is quite reasonable.

As is evident from Fig. 3.17, the excited state absorption at the probe laser wavelength has a substantial effect on the performance of a dye laser amplifier, whereas the excited state absorption at the pump laser wavelength has a rela-

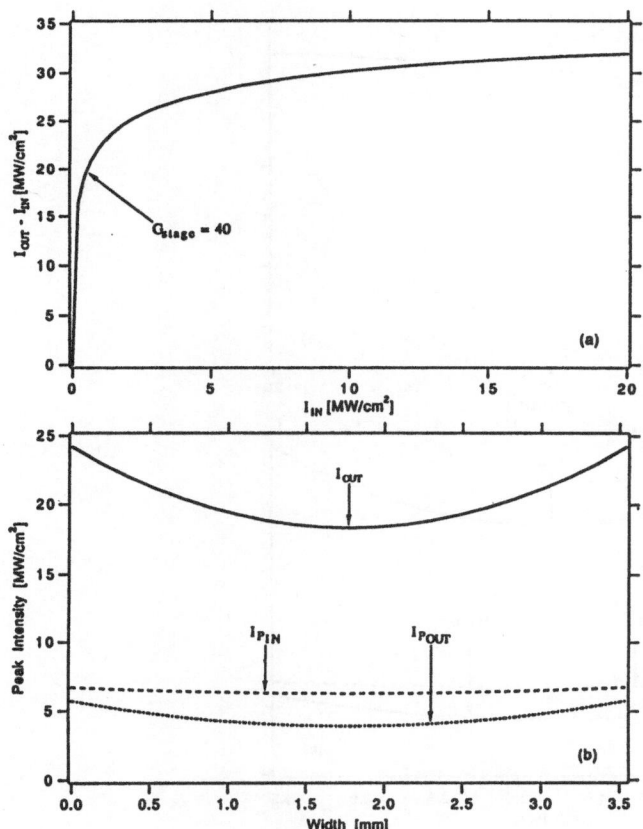

Fig. 3.21. Design calculations of a double-side-pumped dye laser amplifier; (a) probe laser extraction intensity vs. probe laser input intensity with no excited state absorption losses, (b) corresponding intensity profiles of pump and probe laser beams across input and output ends of the amplifier

tively minor one (except at very high intensities). To correct (3.72) for excited state absorption at the probe laser wavelength is quite straightforward. Equation (3.57) is modified to include excited state absorption so that

$$\frac{dI}{dy} = n_1(\sigma_e - \sigma^L_{12})I \ , \tag{3.74}$$

and (3.60) becomes

$$\ln \frac{I_{out}}{I_{in}} = (1 - R_e)G \tag{3.75}$$

where R_e is defined by (3.21b). Following the same derivation as that of (3.72)

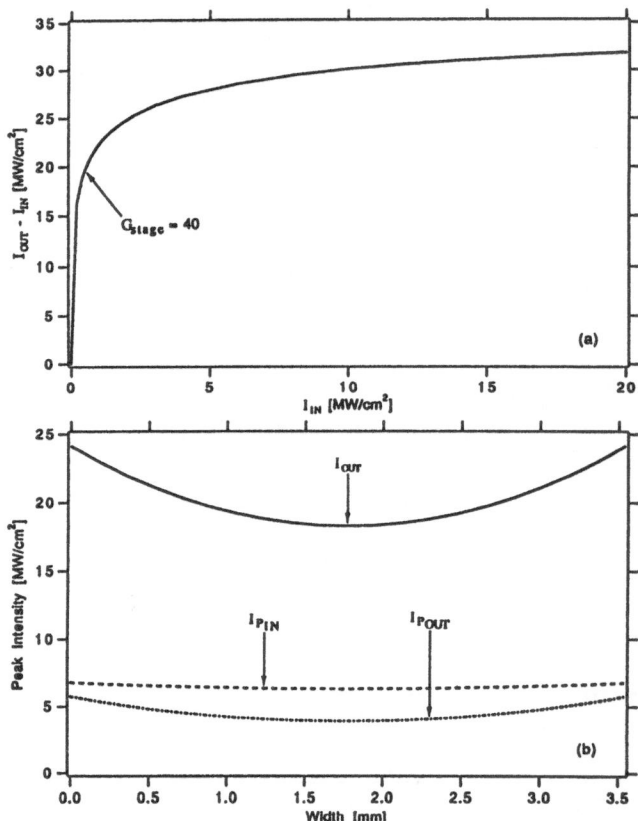

Fig. 3.22. Design calculations of a double-side-pumped dye laser amplifier; (a) probe laser extraction intensity vs. probe laser input intensity with excited state absorption losses, (b) corresponding intensity profiles of pump and probe laser beams across input and output ends of the amplifier

gives

$$\ln \frac{\phi_{\text{out}}}{\phi_{\text{in}}} + (1 - R_G)(\phi_{\text{out}} - \phi_{\text{in}}) = (1 - R_e)G_0 \tag{3.76}$$

for the small signal gain equation. Repeating the calculation for the amplifier design with $R_e = 0.2$ the estimated length becomes $L = 2.32$ cm which is 20% longer than in the case where $R_e = 0$. The scaling of amplifier length with excited state absorption loss is evident from inspection of (3.76). Again the results are checked with the numerical integration of (3.19, 20) with the new design and optical parameters. The gain curve from the calculation is shown in Fig. 3.22a. At the input intensity, $I_{\text{in}} = 0.5$ MW/cm^2, the calculated output intensity is

$I_{out} = 20.2 \, \text{MW}/\text{cm}^2$ which is again in close agreement with the design estimate. The spatial profiles of the pump and probe beams at the ends of the amplifier are shown in Fig. 3.22b. The output beam of the amplifier is reasonably uniform. For more accurate amplifier design, the procedure outlined here should be repeated until all the specifications of the amplifier are met.

3.7 Summary

This chapter has presented an alternative method for experimentally determining the fundamental active media parameters important for predicting dye laser performance. This technique employs standard laboratory instrumentation and eliminates the need for both fluorescence lifetime and quantum yield measurements. The resulting measurements have the potential for being both more accurate and more easily accomplished than the independent parameter measurements commonly used.

Additionally, the formalism developed to determine the active media parameters has been applied to predicting dye laser amplifier performance. A system design strategy is presented that employs a simplified design analysis to estimate an appropriate architecture followed by a more complete analysis to predict final performance.

References

3.1   R.S. Taylor, P.B. Corkum: Appl. Phys. **B26**, 31 (1981)
3.2   P. Mazzinghi, V. Uivano, P. Burlamacchi: Appl. Opt. **22**, 3335 (1983)
3.3   P.N. Everett, H.R. Aldag, J.J. Ehrlich, G.S. Janes, D.E. Klimek, F.M. Landers, D.P. Pacheco: Appl. Opt. **25**, 2124 (1986)
3.4   F.P. Schäfer: Principles of dye laser operation, in *Dye Lasers*, ed. by F.P. Schäfer, Topics in Appl. Phys., Vol. 1 (Springer, Berlin, Heidelberg 1973)
3.5   O. Teschke, A. Dienes, J.R. Whinnery: IEEE J. Quantum Electron. **QE-12**, 383 (1976)
3.6   R.S. Becker: *Theory and Interpretation of Fluorescence and Phosphorescence* (Wiley, New York 1969) pp. 80–82
3.7   J.N. Demas, G.A. Crosby: J. Phys. Chem. **75**, 991 (1971)
3.8   K.H. Drexhage: Structure and properties of laser dyes, in *Dye Lasers*, ed. by F.P. Schäfer, Topics in Appl. Phys., Vol. 1 (Springer, Berlin, Heidelberg 1973)
3.9   H.W. Furumoto, H.L. Ceccon: IEEE. J. Quantum Electron. **QE-6**, 262 (1970)
3.10   A.N. Fletcher, D.E. Bliss, J.M. Kauffman: Optics Commun. **47**, 57 (1983)
3.11   T.G. Pavlopoulos, J.H. Boyer, I.R. Politzer, C.M. Lau: Optics Commun. **64**, 367 (1987)
3.12   T.G. Pavlopoulos, M. Shah, J.H. Boyer: Optics Commun. **70**, 425 (1989)
3.13   O. Uchino, T. Mizunami, M. Maeda, Y. Miyazoe: Appl. Phys. **19**, 35 (1979)
3.14   P. Cassard, R.S. Taylor, P.B. Corkum, A.J. Alcock: Optics Commun. **38**, 131 (1981)
3.15   H. Telle, W. Hueffer, D. Basting: Optics Commun. **38**, 402 (1981)
3.16   P. Cassard, P.B. Corkum, A.J. Alcock: Appl. Phys. **25**, 17 (1981)
3.17   F.P. Schäfer: Laser Chem. **3**, 265 (1983)

3.18   V.S. Antonov, K.L. Hohla: Appl. Phys. **B30**, 109 (1983)
3.19   V.S. Antonov, K.L. Hohla: Appl. Phys. **B32**, 9 (1983)
3.20   M. Rinke, H. Guesten, H.J. Ache: J. Phys. Chem. **90**, 2661 (1986)
3.21   C.H. Chen, J.L. Fox, F.J. Duarte, J.J. Ehrlich: Appl. Optics **27**, 443 (1988)
3.22   M.R. Padhye, T.S. Varadarajan, A.V. Deshpande: Spectrosc. Lett. **17**, 369 (1984)
3.23   V.I. Tomin, A.J. Alcock, W.J. Sarjeant, K.E. Leopold: Optics. Commun. **26**, 396 (1978)
3.24   M. Asscher, Y. Haas: Appl. Phys. **20**, 291 (1979)
3.25   W. Zapka: Appl. Phys. **20**, 283 (1979)
3.26   F.-G. Zhang, F.P. Schäfer: Appl. Phys. **26**, 211 (1981)
3.27   A.N. Fletcher, R.H. Knipe, M.E. Pietrak: Appl. Phys. **B27**, 93 (1982)
3.28   B. Kopainsky, P. Qiu, W. Kaiser: Appl. Phys. **B29**, 15 (1982)
3.29   A.N. Fletcher, R.H. Knipe: Appl. Phys. **B29**, 139 (1982)
3.30   A.N. Fletcher: Appl. Phys. **B31**, 19 (1983)
3.31   R.H. Knipe, A.N. Fletcher: J. Photochem. **23**, 117 (1983)
3.32   RJ. von Trebra, T.H. Koch: Appl. Phys. Lett. **42**, 129 (1983)
3.33   T. Mukherjee, K.N. Rao, J.P. Mittal: Indian J. Chem. **25A**, 509 (1986)
3.34   A.N. Fletcher, M.E. Pietrak, D.E. Bliss: Appl. Phys. **B42**, 79 (1987)
3.35   J.M. Kauffman, J.H. Bentley: Laser Chem. **8**, 49 (1988)
3.36   U. Ganiel, A. Hardy, G. Neumann, D. Treves: IEEE J. Quantum Electron. **QE-11**, 881 (1975)
3.37   T. Efthimiopoulos, B. K. Garside: Can. J. Phys. **59**, 820 (1981)
3.38   I.A. McIntyre, M.H. Dunn: Optics Commun. **50**, 169 (1984)
3.39   A.A. Hnilo, O.E. Martinez, E.J. Quel: IEEE J. Quantum Electron. **QE-22**, 20 (1986)
3.40   B.B. Snavely: Proc. IEEE **57**, 1374 (1969)
3.41   O.G. Peterson, W.C. McColgin, J.H. Eberly: Phys. Lett. **29A**, 399 (1969)
3.42   A.V. Buettner, B.B. Snavely, O.G. Peterson: *Molecular Luminescence* ed. by E. Lim (Benjamin, New York 1969) p. 403
3.43   C. Koepke: Acta Phys. Pol. **A66**, 741 (1984)
3.44   J.P. Webb, W.C. McColgin, O.G. Peterson, D.L. Stockman, J.H. Eberly: Chem. Phys. **53**, 4227 (1970)
3.45   O.G. Peterson, J.P. Webb, W.C. McColgin, J.H. Eberly: J. Appl. Phys. **42**, 1917 (1971)
3.46   T.G. Pavlopoulos, J.H. Boyer, I.R. Plitzer, C.M. Lau: J. Appl. Phys. **60**, 4028 (1986)
3.47   T.G. Pavlopoulos, D.J. Golich: J. Appl. Phys. **64**, 521 (1988)
3.48   T.G. Pavlopoulos, D.J. Golich: J. Appl. Phys. **67**, 1203 (1990)
3.49   D.J. McClure: J. Chem. Phys. **19**, 670 (1951)
3.50   J.B. Marling, D.W. Gregg, L. Wood: Appl. Phys. Lett. **17**, 527 (1970)
3.51   A.E. Siegman, D.W. Phillion, D.J. Kuizenga: Appl. Phys. Lett. **21**, 345 (1972)
3.52   D.E. Klimek: Appl. Phys. **B34**, 83 (1984)
3.53   A.E. Siegman: *Lasers* (University Science Books, Mill Valley 1986) p. 294
3.54   E. Sahar, D. Treves: IEEE J. Quantum Electron. **QE-13**, 962 (1977)
3.55   R.S. Taylor, S. Mihailov: Appl. Phys. **B38**, 131 (1985)
3.56   C. Lin: IEEE J. Quantum Electron. **QE-11**, 602 (1975)
3.57   H. Uchiki, M. Yoshizawa, T. Kobayashi: IEEE J. Quantum Electron. **QE-19**, 551 (1983)
3.58   R.G. Morton, M.E. Mack, I. Itzkan: Appl. Optics **17**, 3268 (1978)
3.59   F.P. Schäfer, L. Wenchong, S. Szatmari: Appl. Phys. **B32**, 123 (1983)
3.60   S. Szatmari, F.P. Schäfer: Appl. Phys. **B33**, 95 (1984)
3.61   P.H. Chiu, S. Hsu, S.J.C. Box, H.-S. Kwok: IEEE J. Quantum Electron. **QE-20**, 652 (1984)
3.62   Z. Bor, B. Racz: Appl. Optics **24**, 1910 (1985)
3.63   P. Simon, J. Klebniczki, G. Szabo: Optics Commun. **56**, 359 (1986)
3.64   Z. Bor, F. Raksi, G. Kovacs, B. Racz: Appl. Spectrosc. **42**, 583 (1988)
3.65   I.H. Munro, I.A. Ramsay: J. Sci. Instr. **1**, 147 (1968)
3.66   A.E.W. Knight, B.K. Selinger: Spectrochim. Acta **27A**, 1223 (1971)
3.67   U.P. Wild, A.R. Holzwarth, H.P. Good: Rev. Seci. Instrum. **48**, 1621 (1977)
3.68   M.S. Murillo: Coupled cavity quenched dye laser and analysis for fluorescence lifetime measurements; Los Alamos National Laboratory report no. LA-UR-90-2082

3.69   J.B. Birks: *Photophysics of Aromatic Molecules* (Wiley, New York 1970) Chap. 4
3.70   P.R. Hammond: J. Chem. Phys. **70**, 3884 (1979)
3.71   I. Wieder: Appl. Phys. Lett. **21**, 318 (1972)
3.72   L. Huff, L.G. DeShazer: J. Opt. Soc. Am. **60**, 157 (1970)
3.73   G. Haag, G. Marowsky: IEEE J. Quantum Electron. **QE-16**, 890 (1980)
3.74   M. Watanabe, A. Endoh, S. Watanabe: Proceedings of the International Conference on Lasers, ed. by F.J. Duarte (STS Press, McLean, VA 1988)
3.75   E. Sahar, I. Wieder: Chem. Phys. Lett. **23**, 518 (1973)
3.76   E. Sahar, I. Wieder: IEEE J. Quantum Electron. **QE-10**, 612 (1974)
3.77   J. Shah, R.F. Leheny: Appl. Phys. Lett. **24**, 562 (1974)
3.78   J. Faure, J.-P. Fouassier, D.-J. Lougnot: Phys. Lett. **50A**, 319 (1974)
3.79   J. Langelaar: Appl. Phys. **6**, 61 (1975)
3.80   O. Teschke, A. Dienes: Appl. Phys. Lett. **26**, 13 (1975)
3.81   R.F. Leheny, J. Shah: IEEE J. Quantum Electron. **QE-11**, 70 (1975)
3.82   C.D. Decker: Appl. Phys. Lett. **27**, 607 (1975)
3.83   G. Dolan, C.R. Goldschmidt: Chem. Phys. Lett. **39**, 320 (1976)
3.84   A. Penzkofer, W. Falkenstein, W. Kaiser: Chem. Phys. Lett. **44**, 82 (1976)
3.85   G.A. Abakumov, S.A. Vorob'ev, A.P. Simonov: Sov. J. Quantum Electron. **7**, 1094 (1977)
3.86   A. Mueller, J. Schulz-Hennig, H. Tashiro: Appl. Phys. **12**, 333 (1977)
3.87   W. Falkenstein, A. Penzkofer, W. Kaiser: **27**, 151 (1978)
3.88   P.R. Hammond: IEEE J. Quantum Electron. **QE-15**, 624 (1979)
3.89   W. Zapka, F.P. Schäfer: Appl. Phys. **20**, 287 (1979)
3.90   D. Magde, S.T. Gaffney, B.F. Campbell: IEEE J. Quantum Electron. **QE-17**, 489 (1981)
3.91   M. Hercher: Appl. Optics: **6**, 947 (1967)
3.92   E.A. Stappaerts, W.H. Long, Jr.: Optics Lett. **3**, 226 (1978)
3.93   P.R. Hammond: Appl. Opt. **18**, 536 (1979)
3.94   C.S. Sexton, W.H. Steier: IEEE J. Quantum Electron. **QE-20**, 1114 (1984)
3.95   CH.G. Christov, I.V. Tomov: Opt. Quantum Electron. **18**, 137 (1986)
3.96   S. Szatmari, B. Racz: Opt. Quantum Electron. **19**, 20 (1987)
3.97   N. Michailov, T. Deligeorgiev, V. Petrov, I. Tomov: Opt. Commun. **70**, 137 (1989)
3.98   P. Cassard, P.B. Corkum, A.J. Alcock: Appl. Phys. **25**, 17 (1981)
3.99   M. Faraggi, P. Peretz, I. Rosenthal, D. Weinraub: Chem. Phys. Lett. **103**, 310 (1984)
3.100   I.A. McIntyre, M.H. Dunn: J. Phys. E: Sci. Instrum. **18**, 19 (1985)
3.101   S. Dellonte, E. Gardini, F. Barigelletti, G. Orlandi: Chem. Phys. Lett. **49**, 596 (1977)
3.102   G. Orlandi, L. Flamigni: Radiat. Phys. Chem. **21**, 113 (1983)
3.103   W.H. Press, B.P. Flannery, S.A. Teukolsky, W.T. Vettering: *Numerical Recipes* (Cambridge University Press, Cambridge 1986) Chap. 14
3.104   E.H. Gilmore, G.E. Gibson, D.S. McClure: J. Chem. Phys. **20**, 829 (1952)
3.105   W.H. Melhuish: J. Phys. Chem. **65**, 229 (1961)
3.106   P.G. Seybold, M. Gouterman, J. Callis: Photochem. Photobiol. **9**, 229 (1969)
3.107   O.G. Peterson: Dye lasers, in *Methods in Experimental Physics* ed. by C.L. Tang, Vol. 15A (Academic, New York 1979)

4. Large-Scale, Excimer-Laser-Pumped Dye Lasers

C. Tallman and R. Tennant

With 28 Figures

Successful design of efficient, long-life, high-intensity dye lasers is a challenging balance of many critical factors. Designing for high power is particularly challenging because of the very high saturation intensity of most dyes, requiring pumping intensities of the order of 10 MW/cm^2. Efficient power extraction often requires operation near the threshold of optical damage for the optical components and potential onset of photochemistry in the dye and solvent. This chapter will address these critical factors.

4.1 Basics

The advent of XeCl excimer laser pumping of dye lasers at 308 nm has opened the door to very high conversion efficiencies and in some ways has simplified the design over flashlamp-pumped dye lasers. We will discuss the advantages of excimer pumping, in particular why the XeCl laser is best. We also will discuss in depth the reliability of the excimer lasers and will assess the expected performance of the full dye laser system.

4.1.1 Design Configuration

The basic design approach addressed will be limited to the use of two-sided transverse excimer laser pumping of a dye cell. The designs considered in this chapter follow those analyzed in Chap. 3. The reader is referred to that chapter for guidance in sizing the dye cell, based on dye characterizing parameters, and for the required performance specifications.

The dye is dissolved in a fluid solvent and is circulated through the dye cell. The optical configurations will be stable or unstable resonators with high-power, one-pass amplifier designs. Line narrowing and tuning is accomplished by introducing dispersive elements in the optical cavity or etalons. Chapter 2 gives specific guidance on the optical cavity and optical components. For high-repetition rate, the design will allow the dye to flow orthogonal to the pump and to the dye beam as shown in Fig. 4.1.

We will methodically consider each component in this design and the interactive coupling of these components on system performance. Firstly, let us look into the design details of the dye cell itself.

Springer Series in Optical Sciences, Vol. 65
High-Power Dye Lasers Editor: Francisco J. Duarte
© Springer-Verlag Berlin Heidelberg 1991

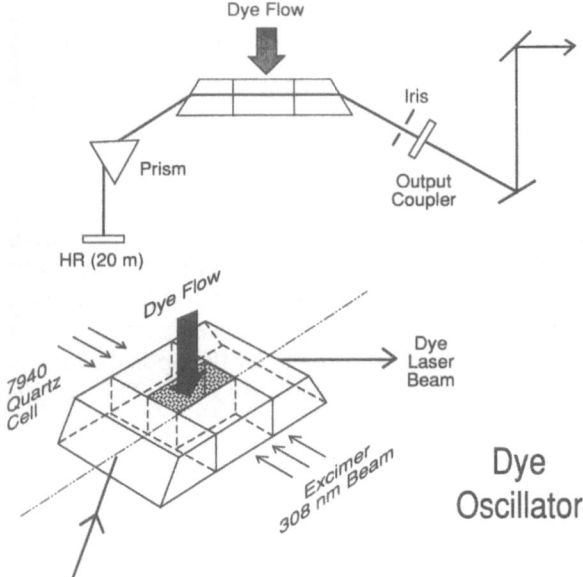

Fig. 4.1. Basic dye laser configuration showing the dye cell, excimer laser transverse pump and dye fluid flow (from [4.23])

4.1.2 Dye-Cell Design

Successful dye-cell designs for high-power excimer pumping use transverse pumping where the pump beam, dye extraction beam and flow directions are mutually orthogonal. *Hargrove* et al. [4.1] and *Klick* [4.2] demonstrated the advantages of two-sided transverse pumping compared with longitudinal pumping for high-repetition-rate dye laser systems. This design provides the flexibility of the extraction efficiency by changing the aspect ratio (l/d). For high-repetition-rate, high-power, dye-cell designs this configuration allows minimum exposure time of the dye to the pump beam. An example of this design, used for a high-power excimer laser dye system at Los Alamos, is shown in Fig. 4.2 [4.3].

4.1.3 Flow Requirements

Ideally, the dye fluid exposed to the excimer pump beam should be exchanged in the cell at least twice per excimer laser pulse. It is important to minimize the induced index gradients in the dye cell to achieve high beam quality. The excimer laser beam is absorbed in the dye, and at least half the energy results in heating the dye and solvent. Temperature gradients will produce index gradients and thus dye beam diffraction. Some solvents have larger dn/dT than others, and thus the flow rate is somewhat dependent upon the solvent's optical properties. Furthermore, the collimated excimer laser beam will not be absorbed uniformly

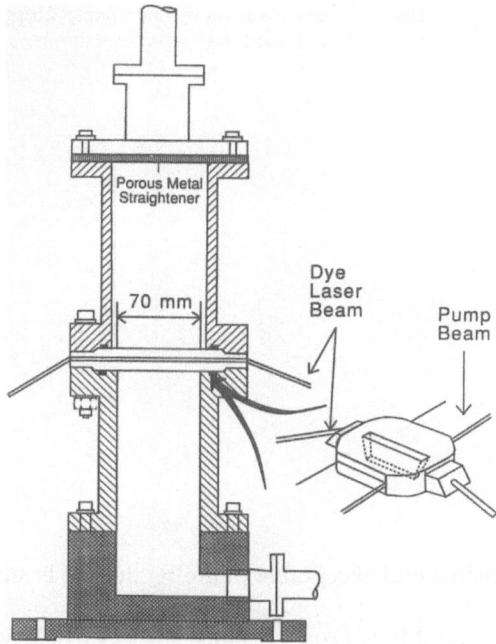

Fig. 4.2. A 7940 Corning Fused Silica monolithic dye cell, showing the excimer pump windows, dye cell brewster windows and flow passage

throughout the cell width. A cell, as shown in Fig. 4.2, with a plane window will exhibit an absorption curve through the cell width as shown in Fig. 4.3.

Choosing the cell width and dye concentration to optimize the pumping uniformity will still result in higher energy deposition in the fluid at the pump windows. The dye flow rate is chosen to flush out the optically pumped dye after each excimer laser pulse. The velocity profile is not uniform through the flow passage. A boundary layer of very slow-moving dye exists at the window surface, so the flow rates must be sufficient to remove this slow-moving layer. This fluid is exposed to the highest pumping fluence. Clearly then the volume flow rate must exceed the average volume flow rate by a factor dependent upon the beam quality and also upon optical damage thresholds

$$V_g > \frac{v}{T} , \qquad (4.1)$$

where V_g is the average volume flow rate, v is the cell volume being optically pumped and T is the pulse repetition period.

Typically the number of dye volumes replaced per shot is at least two or more. The design of the flow passage can have a major effect on both the

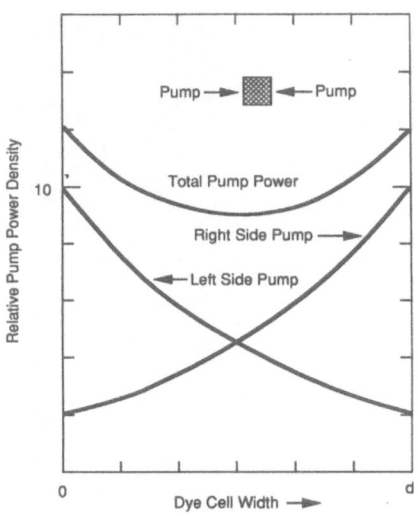

Fig. 4.3. Absorption of excimer pump energy across the cell width for two-sided pumping

minimization of induced index gradients and acceptable pumping fluence levels without onset of damage.

Flow calculations for nonsymmetrical flow cross sections, such as dye flow cells, are very complex. Exact treatment for three-dimensional flow in the corners between the windows is not readily tractable. Some experimental work must be done on a flow model. An experimental off-line facility to flow test new designs and to observe the flow patterns in these complex geometrical flow passages is highly desirable.

4.1.4 Dye Flow Loop

A high clearing ratio is necessary to have extended window and dye lifetimes. The upper limit is set by the fluid pump, available pressure head, power, size of the flow hardware and pressure drop across filters in the flow system. The higher the repetition rate and energy per pulse required from the dye laser, the higher the required flow rates. Pump-produced vibrations or flow surges can generate observable index gradients. The flow design may require incorporation of hydraulic reservoirs, flow smoothing methods and careful isolation of flow-induced vibration to achieve the required beam quality. Dye flow-handling system considerations become a major concern for dye systems designed to operate at high repetition rates reaching 500 Hz–1 kHz. Although high-power commercial lasers are now available for operation at 500 Hz and will soon be available to 1 kHz, designing flow systems for these high repetition rates may not be practical. Alternative approaches using multiplexing of several lower repetition rate dye lasers, for example, may have significant advantages when considering the flow handling and safety problems of high-repetition-rate high-power systems.

4.2 Dye and Solvent Considerations

4.2.1 Dye Selection

Selection of a suitable dye involves much more than finding a dye with absorption at the pump wavelength and an emission peak at the desired wavelength. Of major consideration is the dye's efficiency and stability. A rather large number of dyes have been evaluated for XeCl excimer pumping, covering the spectral range from the near UV of 340 nm to the IR at over 1 micron [4.2, 4.4–6].

4.2.2 Efficiency

Although much has been reported on efficiency and stability, it is prudent that we take a moment and clearly review the definitions of efficiency and lifetime and the factors that can cloud the issues in understanding the reported results. Some very excellent comparative measurements showing relative efficiencies of dyes, dye mixes, solvents and solvent mixes have been made in various oscillator configurations. These results are dependent upon the oscillator design, cavity losses and pumping intensities, which makes comparisons difficult. Recent work by *Antonov* at the Institute of Spectroscopy of the Academy of Science, Podol'skii, USSR, and *Brackmann* of Lambda Physik, Gottingen, Germany, [4.7] have evaluated dyes over the spectral range of IR to UV and have reported energy conversion efficiencies from 18% for the IR dyes to as high as 39% for Coumarin 120 in the visible. They chose to make all their measurements using a saturated dye laser amplifier configuration. This procedure, when properly checked, gives the most unambiguous results. They clearly defined the efficiencies being measured. The system efficiency is simply the total dye-laser output energy divided by the total input pump energy

$$n = E_{\text{dye-laser}}/E_{\text{pump}} \ . \tag{4.2}$$

This equation is to be distinguished from quantum efficiency, which is

$$n_{\text{q}} = \frac{N_{\text{Lp}}}{N_{\text{Pp}}} \ , \tag{4.3}$$

where N_{Lp} is the number of output laser photons and N_{Pp} is the number of input pump photons. The quantum efficiency is to be distinguished from "fluorescence efficiency," which includes all the spontaneously emitted photons. To be sure that the output does not include the spontaneous emission, the dye amplifier is driven well into saturation for the efficiency measurements.

Oscillator laser configurations include uncertainties, such as threshold and various absorption losses that are difficult to treat quantitatively, whereas the saturated amplifier measurement gives a more direct answer. They verified that their measurements were conducted well into saturation. The saturation power

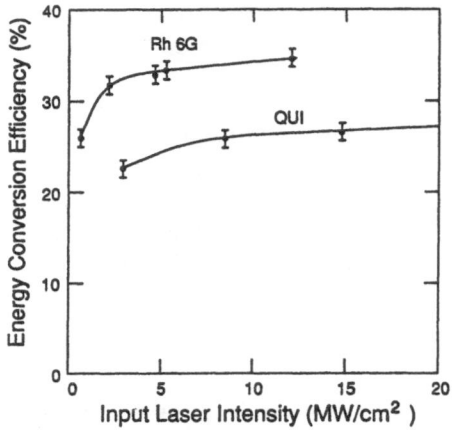

Fig. 4.4. Energy conversion efficiency vs. input laser intensity for Rh 6G and QUI (from [4.7])

density is

$$I_S = \frac{hv}{tq} , \tag{4.4}$$

where I_S is the saturation power density, h is Plank's constant, v is the frequency of the light, t is the lifetime of the inversion and q is the cross section for stimulated emission. This value is typically 2–5 MW/cm^2 for most dyes in the emission band. Measurements were made on all the dyes tested. A typical plot of conversion efficiency vs. input laser power density is shown for Rh 6G and QUI dyes in Fig. 4.4. The curves are essentially flat with little change in conversion efficiencies for input intensities exceeding 3–4 MW/cm^2. Operating the amplifier well into saturation will insure that the output energy then is simply

$$E_{out} = E_{in} + GE_P , \tag{4.5}$$

where E_{out} and E_{in} are the output and input energy and G is the conversion efficiency. The input beam must sweep out the full pumped volume for this expression to be valid.

High pump intensities can begin to pump singlet–singlet and in some cases even triplet–triplet absorption and possibly introduce significant destruction of the dye molecule and/or solvent. Research at the National Research Council of Canada by *Cassard* et al. [4.8] showed that PTP-dioxane exhibited a rapid decrease in the conversion efficiency when pumped at a power density exceeding 9–10 MW/cm^2. They found that for the majority of dyes, there is a maximum pumping intensity of about 10 MW/cm^2. Antonov and Brackmann showed that QUI drops in conversion efficiency when the pumping intensity exceeds 10 MW/cm^2; however, at 20 MW/cm^2 the efficiency is only reduced by 10%. Our experiments at Los Alamos with TBS-dioxane indicate that pumping intensities in excess of 20 MW/cm^2 are possible without significant loss in

Table 4.1. Conversion and quantum efficiencies for selected dyes

Dye	Peak [nm]	Solvent	Energy conversion efficiency [%]	Quantum efficiency [%]
BiBuQ	385	dioxane	30.0	37.5
		cyclohex	30.0	37.5
QUI	386	dioxane	27.4	34.3
Popop	417	dioxane	21.0	28.4
Coumarin 47	458	methanol	34.0	50.6
Coumarin 120	457	methanol	34.0	60.1
Rhodamine 6G	581	methanol	35.0	66.1
DCM	658	DMSO	21.0	44.9
Pyridine 1	710	DMSO	21.0	48.4
Rhodamine 700	722	methanol	28.0	65.6
Pyridine 2	740	DMSO	23.0	55.3
Styryl 9	840	DMSO	18.5	50.0

efficiency. (However we did observe very dramatic solvent photochemistry in cyclohexane with consequential damage to the dye cell, which will be discussed later). With some dyes and solvents, this damage can be a major factor, affecting not only the conversion efficiency but also the dye lifetime. A good procedure would be to check any new selected dye for sensitivity to pump intensity before putting the dye into service.

The conversion efficiencies measured by Antonov and Brackmann are surprisingly high, and the quantum efficiencies for the IR dyes are also very high. Table 4.1 [4.7] shows their results.

These outstanding efficiencies are near the upper limit as one can expect; however, it should be understood that these data were taken using a single-sided pumped dye cell. The pumping was not uniform across the excited volume nor was special care taken to homogenize the pump beam in height or width. Optimizing the pumping uniformity, selecting a more optimum solvent and varying the concentration might yield even higher conversion efficiencies. Recent molecular engineering of dyes tailored so that their absorption peaks near the 308 nm pump wavelength and designing for high efficiency and long lifetime have shown remarkable performances [4.9, 10]. Much has yet to be done in this area, and we can expect even higher efficiencies and stabilities for new dyes designed specifically for XeCl pumping.

4.2.3 Dye Stability

Photochemical stability is one of the most important considerations in selecting a dye and has been a major concern for years with flashlamp-pumped dyes. For excimer-laser-pumped dyes, the problem is less complex; nonetheless, dye stability is a major factor in selecting dyes for applications that require long-life operation with minimum maintenance. Considerable insight into dye and solvent stability can be gained from these earlier studies and from experiments with flashlamp pumping. *Fletcher* et al. [4.11, 12] produced a very detailed and valuable nine-part study on laser dye stability that spans a decade of work. *Knipe* and *Fletcher* [4.13] also made a theoretical study of the effects of photo-product absorption buildup. They demonstrated that, in some optical cavities, absorption at the laser wavelength of 1% is sufficient to reduce the laser output by 50%.

Dye laser output degradation caused by photo disintegration of the dye molecule and subsequent reactions of the photo-products with the solvent or with impurities is a very complex subject and is not fully understood. Catalytic effects can occur at window or hardware surfaces, which can promote multi-photon reactions. The as-received dye solvent purity and dye purity can play a major role. It is estimated that most common solvents are saturated with air at concentrations of 10^{-3}–10^{-4} M. In many cases this oxygen contaminant reacts with the solvent and the dye under intense UV pumping. The photochemistry may not always cause the lasing energy to decrease in a linear or exponential fashion. In the case of Oxazine 750 in DMSO, the output energy actually increased [4.14]. *Antonov* and *Hohla* [4.14] studied some 34 different laser dyes and reported the stability, the form of the energy decrease (linear, exponential, or some combination) and absorption coefficients with which estimates can be made of the energy decrease. The most practical way to report the dye stability is to determine the total pump energy absorbed by the dye solution before the laser output energy drops to 50% of the laser's initial value. The energy can be expressed as number of pump photons, jouls or watt hours. Since a watt hour is 3600 joules and one 308 nm photon has 6.37×10^{-19} joules, conversion is straightforward. The number of molecules is

$$n = \frac{cN_A l}{M_{dye}} ,$$

(4.6)

where n is the number of dye molecules, N_A is Avagadro's number, M_{dye} is the dye molecular weight and l is the dye volume being irradiated in liters.

Good UV laser dyes should sustain several thousand photons/molecule before dropping to half the original energy. Antonov and Hohla report that QUI in dioxane will take 34 000 photons per molecule – a very high value, in fact the highest reported in the dyes studied. For 1 liter dye volume of QUI at the test concentration of 4×10^{-4} M gives 2.4×10^{20} molecules irradiated. For 34 000 photons/molecule exposure, a total of 8.24×10^{24} photons must be used. Since each 308 nm photon has 6.37×10^{-19} joules, the total number of joules needed would be 5.25×10^6, the same as 1458 watt hours. Delivering 0.2 joules/pulse at 10 Hz (2 watts) into the dye solution then would allow 729 hours of operation. Antonov and Hohla point out that one must be sure that the quoted values are the photons actually absorbed by the dye solution and not what is produced by the excimer laser. Corrections for the transport efficiency to the cell, losses at the

Table 4.2. Photochemical stability of selected dyes

Dye	Lifetime [w-h to 50% energy output]
PTP, PBD	1.5
S1, PV1	1.5
BBD, BMSB	50
S3	2
C 120	20
C 47	4
C 307	8
Rh 6G	50
Rh B	10
Rh 101	20
CV	25
DCM	50
Rh 700	25
Oxa 750	12
DMOTC	20
IR 140/4	12

cell window and photons transmitted must be accounted for. Lifetime numbers for most UV dyes in the optimum solvent typically have an order of magnitude less than that quoted for QUI. Radiant Dyes Chemie [4.15] published a lifetime of 4100 photons/molecule or 78×10^4 joules/liter (217 Wh) for their UV laser dye RDC 374 in dioxane. They also indicate that this same dye, when in ethanol solvent, will have a factor of 10 shorter lifetime. *Hohla* [4.16] lists the typical dyes from 340 nm to over 900 nm. Table 4.2 shows the stability of a selected set of dyes in terms of watt hours of excimer pumping to reach 50% dye output energy.

In the wavelength range of 420–560 nm covered by the Coumarins, the lifetimes are very low. Typical lifetimes are only 100 photons/molecule or less [4.14]. We also find that some of the IR dyes have excimer-pumping lifetimes in the range of 100 photons/molecule. Therefore more research needs to be devoted to developing robust dyes in these portions of the spectrum.

4.2.4 Solvent Effects on Dye Stability

The solvent's influence on stability and the dye photo-product interaction can be major contributors to the lifetime of the dye solution. In their studies of aminocoumarins, in particular C 120, *Kunjappu* et al. [4.17] describe how the presence of oxygen accelerates the photodegradation of the dye and the solvents of increasing polarity. They found that, for C 120, the dye degraded more rapidly in solvents of low polarity and that low polarity had a dramatic influence on the photo-decomposition of C 120 with different solvents. The C 120 decomposition rate was about 10 times faster with cyclohexane solvent than with methanol. Comparing decomposition rates for methanol, ethanol, isopropanol and cyclo- hexane showed that the decomposition rates increased in the same order as the dielectric constant for these solvents (Table 4.3).

The degradation time was even faster when the solution was bubbled with air. *Cassard's* extensive study [4.18] of a number of UV dyes in combination with several solvents showed significant differences in lifetimes. They reported their lifetime data in kilojoules per cubic centimeters for a decrease in output of their dye oscillator of $1/e$ and observed a single exponential decay in output, except for PTP-cyclohexane, for all dye solvent combinations. The PTP- cyclohexane exhibited an apparent two-time constant decay indicative of a two- step process. Later, we will describe our findings at Los Alamos on photochem- istry of cyclohexane, which might explain some of their results. Their reported lifetimes can be converted to watt hours or photons per molecule. It is informa- tive to list some of their results in these units (Table 4.4).

Note the difference in efficiency and lifetime for dioxane vs. isopropanol. Comparisons made by P. Cassard for BBD, PBD, QTP, BBQ and PBZO in the above solvents show a similar strong lifetime dependency on the solvent. Some combinations show good lifetimes; for example, they measured 5775

Table 4.3. Decomposition of C 120 in various solvents

Solvent	Dielectric constant	Time of degradation to 50% (argon-bubbled) [min]
Methanol	33.62	53
Ethanol	24.30	36
Isopropanol	18.30	12.5
Cyclohexane	2.02	3.5

Table 4.4. PTP dye stability for various solvents

Dye	Solvent	Concentration [10^{-3} M/l]	Efficiency [%]	Lifetime (1/e) [kJ/cm^3]	p/m (50%)
PTP	cyclohexane	2	23	0.53	480
	n-hexane	2	23	0.32	289
	heptane	2	23	1.7	1525
	ethanol	2	29	0.24	217
	isopropanol	2	17	0.04	36
	methanol	2	19	0.05	45
	acetonitrile	2	20	0.03	27
	dioxane	2	31	2.1	1899
	NMP	2	18	1.3	1178

photons/molecule for BBD in dioxane with an efficiency of 24%. Note that the two references above show shorter lifetimes for dyes using cyclohexane as the solvent.

4.2.5 Solvent Stability

Smith [4.19], together with Craig Jensen at the Los Alamos National Laboratory, did extensive analysis of cyclohexane solvent for TBS (3,5,3,5 Tetra-*t*-Butyl-*p*-sexiphenyl; manufacturer, Lambda Physik) used in a large MOPA dye laser system. To understand which photo-decomposition reactions were occurring during XeCl irradiation of the TBS-cyclohexane dye solutions, they made studies on the photochemistry of cyclohexane alone. They irradiated cyclohexane in a small flow cell and measured the chemical composition of the solvent after irradiation by an excimer laser. Identical samples of cyclohexane were exposed to 500 000 pulses from the laser, with each sample experiencing a different pulse energy. The laser-beam spatial profile was held at a constant homogenized size. They found that several chemical species were formed during the irradiation series, principally cyclohexanol, cyclohexene, cyclohexanone and dicyclohexyl. From several repetitions of these experiments, they surmised that the cyclohexane underwent a simultaneous two-photon absorption of the pump laser radiation, which eliminated a hydrogen atom and, to some extent, a hydrogen molecule. Previous experiments have measured fluorescence from two-photon excited cyclohexane [4.20, 21]. The cyclohexyl radical, which resulted from elimination of the hydrogen atom, then reacted with the oxygen dissolved in the liquid. In static cells the oxygen would be depleted by reaction with the cyclohexyl radical even at low pulse energies, and the production of oxygen reaction products would be independent of the higher pulse energies. In flowing cells where air was continually introduced as a gas purge, the production of cyclohexanol increased with increasing laser pulse energy at a fixed number of pulses. The production of other photochemical products would show similar dependence on increasing pulse energy but at reduced yields.

In an effort to minimize the production of alcohol and ketone, some variations in the environment of the cyclohexane were introduced. When the air purge was replaced with a helium purge, the yield of oxygen-containing compounds decreased as expected and the yield of cyclohexene showed a strong nonlinear dependence on the laser pulse energy. The maximum yield of cyclohexene, which occurs at maximum laser pulse energy, did not change with the change in purge gas and may suggest that its production is by direct elimination of H_2 and not by a radical chain reaction. In another series of experiments a radical inhibitor, BHT, was introduced to both the air and the He-purged flowing cyclohexane. In both cases the production of cyclohexanol decreased and the yield of cyclohexene remained unchanged. The effect of BHT on the performance of a TBS dye laser was not measured.

A series of similar experiments was performed on dioxane [4.22]. Much less photo-chemical reaction was produced over the same laser pulse energy range.

The two-photon absorption cross section of dioxane was 10^{-51} cm^4-s, which was not significant at the laser intensities employed.

In conclusion, these experiments demonstrate some very revealing results. No matter what the mechanism of production, oxidation products form rapidly. All the oxygen is consumed and produces cyclohexanol, which is the first product in potentially a whole series of oxidation products. Unless precautions are taken both to remove and to prevent dissolution of oxygen in the solvent system, oxidation products will be observed. By analogy, it might be possible to extend this result to all hydrocarbon solvent systems.

Note that in many and especially large dye laser systems, such as those considered for industrial applications, eliminating oxygen from the system is extremely difficult. A more practical approach is to find solvent systems that are less susceptible to reaction with the air. In fact, we did just that and switched from cyclohexane to dioxane as the solvent for TBS used in the dye laser facility at Los Alamos. The difference in system performance with dioxane compared with cyclohexane was dramatic and will be discussed in detail in the section on dye-cell window damage.

TBS-dioxane was used successfully at Los Alamos in a large MOPA laser system. This laser operated at up to 250 Hz repetition rate, producing 150 mJ/pulse at 400 nm [4.23]. Despite the contamination of oxygen and chemical erosion products of dioxane with the metal plumbing, the measured lifetime exceeded 4000 photons/molecule, which compares favorably with the highest UV dye lifetimes reported earlier. This lifetime is quite high considering the fact that this measurement was made with the oscillator part of the MOPA and included all the real system contamination.

4.2.6 Dye Handling

Most dyes are to be treated as carcinogenic. However, very few rigorous studies have been made on dyes. The system should be designed for minimum exposure of personnel to the dye and solvents and is easy to accomplish for a fully sealed system. However, the problems arise when the system is being opened or when new dye solutions are made and introduced into the system.

4.2.7 Solvent Handling

The two solvents used in the Los Alamos high-power laser system were cyclohexane and dioxane. They are typical of many of the solvents needed for UV dyes. Cyclohexane is a colorless, water insoluble and highly flammable liquid with a sweet odor. The DOT hazard class of solvent is a flammable liquid; the flash point is $-27°C$ with a lower explosive limit of 1.3% and an upper explosive limit of 8.4%. The threshold limit value is 300 ppm (1050 mg/m^3) as established by the National Institute for Occupational Safety and Health (NIOSH) in 1981. Cyclohexane is absorbed by inhalation, and the vapor is mildly irritating to the mucous membranes. The liquid is a fat solvent and thus irritates the skin. General safety precautions are as follows:

- Keep containers in a well-ventilated location.
- Keep away from sources of ignition; hence, no smoking in areas of use.
- Take precautionary measures against static discharges.
- Avoid skin contact.
- Don't smoke, eat or drink when handling the solvent.
- Keep all containers and solutions tightly closed and away from sparks and open flames, and when being used, respiratory protection by the appropriate chemical mask is recommended.

Dioxane is a colorless, volatile, and very hygroscopic liquid with a slightly aromatic smell. A solvent in the DOT hazard class is a flammable liquid with a flash point of 11°C, a lower explosion limit of 1.97%, and an upper explosion limit of 22.5%. Dioxane is suspected of being a carcinogen and a neoplastic agent. Dioxane may form explosive peroxides; hence, a check for the presence of peroxides is recommended before concentrating. The threshold limit value is 25 ppm (90 mg/m^3). The liquid is painful to the eyes and irritating to the skin on prolonged contact, and it can be absorbed through the skin in toxic amounts. Dioxane is insidious – its vapors have poor warning properties; they are faint and inoffensive. Concentrations in air of 300 ppm cause irritation of the eyes, nose and throat. The vapors can be inhaled in amounts that cause serious systemic injury, particularly to the liver and kidneys.

General safety precautions are as follows:

- Keep containers in a well-ventilated location.
- Keep away from sources of ignition; hence, no smoking in areas of use.
- Take precautionary measures against static discharges.
- Avoid skin contact.
- Do not smoke, eat, or drink when handling the solvent.
- Keep containers and solutions tightly closed and away from sparks and open flames.
- Wear protective clothing during use (the appropriate chemical mask, safety goggles and a face shield, a protective suit and rubber gloves).

4.2.8 Dye Solubility

A major consideration in selecting a dye solvent combination is solubility. For high-power dye laser systems pumped by high-power excimer lasers, the requirements are less demanding. For high-power dye lasers, the dye-laser exit window area is ultimately determined by the maximum fluence sustainable without damage. For example, a 1 J/pulse dye laser would have a beam area of not much less than 1 cm^2 to be conservatively below a 2 J/cm^2 damage threshold, considering that the peak-to-average spatial distribution is about 2 for the Gaussian beam profile. The dye cell then has a 1 cm width, and the excimer laser beam must be absorbed through this width. For a uniform gain profile across the cell width, it is desirable to have each of the two pump beams reach 50% absorption at the mid-point of the dye chamber, which will result in the absorption curves shown in Fig. 4.3. The dye concentration needed for 50% absorption in 5 mm is calculated as follows

$$A = \log_{10}\left(\frac{I_0}{I}\right) \tag{4.7}$$

where A is the chemical absorbance, I_0 is the intensity of the input beam and I is the output intensity, and

$$\frac{I}{I_0} = 10^{-emd} \tag{4.8}$$

where e is the molar decadic extinction coefficient, m is the molar concentration and d is the depth into the solution.

For TBS, e is given by Lambda Physik as $7.5 \times 10^4 \, l \, \text{mole}^{-1} \text{cm}^{-1}$, and for this example d is 0.5 cm and I/I_0 is 0.5

$$m = \left(\frac{1}{ed}\right) \log_{10}\left(\frac{I}{I_0}\right). \tag{4.9}$$

The molar concentration then would be 8×10^{-6} moles/l, and for TBS the molecular weight is 683. This gives a concentration of only 7 mg/l. Solubility of TBS in dioxane was 75 mg/l at 20°C. This exercise illustrates that, for high-power dye lasers where the cell widths are of the order of a centimeter, the concentrations are low and the solubility need not be as high as for small systems.

Solubility levels of 10^{-3} moles/l have been considered a requirement for dye-solvent systems; however, as seen in these examples, this is dependent on the cell design and pumping procedure. Typically smaller cell geometries used in the oscillators require higher concentrations. TBS in dioxane for a 1 mm cell depth for two-sided pumping would require 70 mg/l, which is very near the solubility limit for TBS in dioxane. For small oscillators the energy is much less; the beam size can be reduced without reaching the damage threshold and one-sided pumping is sufficient. Single-sided pumping of a 1 mm cell width for 90% absorption in 1 mm for TBS in dioxane would require only 5 mg/l concentration.

Dye–solvent concentrations that approach the solubility limit can present serious problems. In some cases it is difficult to dissolve the dye. For example, dissolving TBS in cyclohexane required heating the solvent to 80°C (boiling point) and was accomplished using a reflux system in a fume hood. The flash point for cyclohexane is -18°C; thus, this procedure presents a serious fire hazard. The TBS, once dissolved at concentrations of 25 mg/l, were reasonably stable, but we observed some precipitation in static storage.

4.3 Optical Damage

Optical damage to the dye-cell windows occurs in most dye lasers at some point because the intensities of the lasing beam and pump beams are often near the damage threshold of the windows. For reliable long-life designs, consideration must be given to the causes of damage and operational limits. Despite years of testing optical components, much remains to be learned about the causes of damage. Damage thresholds are dependent on the window material, surface quality, cleaning methods, window surface and liquid interface, heat transfer, flow conditions, dissolved gases in the liquid, laser wavelength, power density, fluence and even the area under radiation. We will explore the various mechanisms for low-damage thresholds for dye-cell windows compared with those found on clean, bare quartz windows in air.

In practice, for a given pump fluence, liquid–solid surface interfaces suffer catastrophic damage at much lower fluxes and in far fewer total pulses than do gas–solid interfaces. In general, dye laser output energy densities seldom reach values sufficient to cause direct material damage based only on the properties of the optical material and the laser wavelength. One of the most important causes for damage at lower thresholds is deposition of particulates on the windows.

4.3.1 Optical Damage Studies

Optical damage studies were done as a part of the dye-laser development at Los Alamos [4.24]. The experiment utilized a simple flow loop and a single-pass optical cavity. The window materials investigated were quartz, sapphire and magnesium fluoride. Two solvents were tested: cyclohexane and dioxane. No dye was added to the solvents. The optic surfaces were viewed off-axis with video equipment, which had a 5–10 micron resolution. Figure 4.5 shows a schematic drawing of the test setup.

For the tests with cyclohexane, damage, appearing as the deposition of carbon or carbon compounds, was produced at fluence levels as low as 0.4 J/cm^2. In contrast, thresholds for bare quartz tested in air range from 4–10 J/cm^2, and carbon deposits are not part of the damage mechanism.

The damage data for cyclohexane (Fig. 4.6) indicate asymptotic behavior at threshold. At fluences above threshold, all damage is delayed and occurs at approximately 10^5 shots. Tests at various pulse-repetition rates (100–200 Hz), solvent-flow rates (1–0.25 l/s) and clearing ratios (1–3) reveal no significant dependence of damage on any of these parameters.

In an attempt to increase the window damage threshold when in contact with cyclohexane, we tested two other window materials. Sapphire windows exhibited threshold damage as carbon deposits at 1.3 J/cm^2. Sapphire windows tested in air produced a damage threshold of 2 J/cm^2. A magnesium fluoride window could not be damaged in 10^6 shots at fluences up to 3.3 J/cm^2, which was the maximum fluence deliverable by the test equipment at that time.

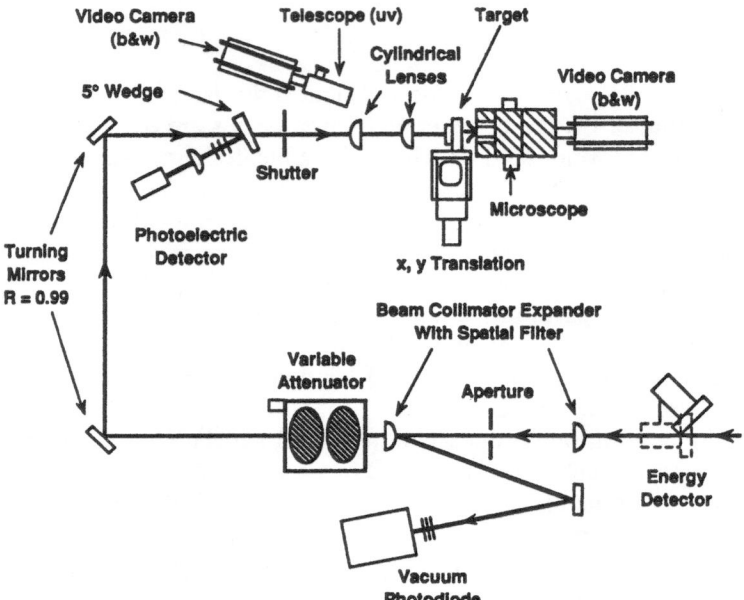

Fig. 4.5. Optical element damage facility test set up

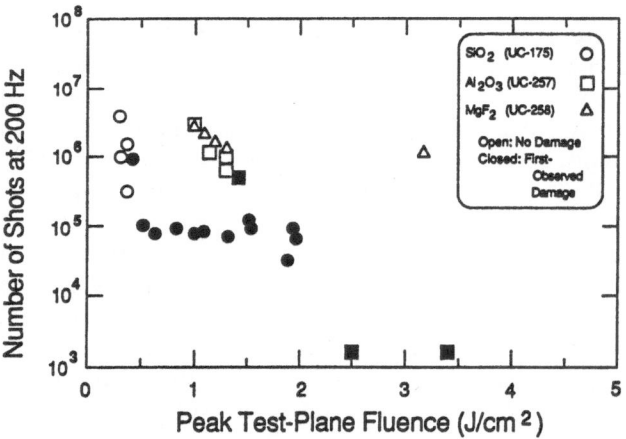

Fig. 4.6. Damage threshold measurement data at 308 nm for various window materials when in contact with flowing cyclohexane

Previous tests of magnesium fluoride in air at 248 and 351 nm produced thresholds of 19 and 20 J/cm², respectively. Dielectric coatings (Al_2O_3) were tested as possible damage-resistant barriers between the solvent and the quartz. Thicker coatings (10 000 Å) can improve damage threshold nearly 1.5 times, whereas thinner coatings provide little or no improvement. These combined results imply that the window is not simply a passive surface that collects the products of photo-dissociation but is an active participant in the photochemical reactions. However, note that quartz has the lowest thermal conductivity and magnesium fluoride the highest; hence, the damage mechanism may be affected in some way by the local heat transfer.

A major improvement in the damage threshold of quartz occurred when the solvent dioxane was used. As seen in Fig. 4.7, the thresholds with dioxane exceed those measured in cyclohexane by an order of magnitude. No carbon deposits were observed on the cell windows. Damage appeared to be a surface-roughening process similar to that seen on samples tested in air. Although companion samples tested in air and in dioxane produce nearly the same damage thresholds ($8–10 \ J/cm^2$), those tested in dioxane appeared to produce delayed damage. Cyclohexane is a suitable solvent for nonpolar laser dyes. Its high optical transparency in the UV allows application in dye lasers pumped below 300 nm. However, the photochemical stability is poor, as discussed earlier, especially under high-power conditions. The general conclusion is that the cyclohexane was being oxidized to products that were acidic in nature, highly hydroxylated, and insoluble in cyclohexane.

Cyclohexanol and dicyclohexyl were the major reaction products observed. Hence, the reaction products would precipitate from the solution and stick to the surfaces of the dye cell and the flow system. These compounds can be strong metal complexants and may have been dissolving metals from the surface of the flow system. Comparison of studies with quartz, sapphire and magnesium fluoride produced no significant differences in the production rate or quantity of reaction products observed.

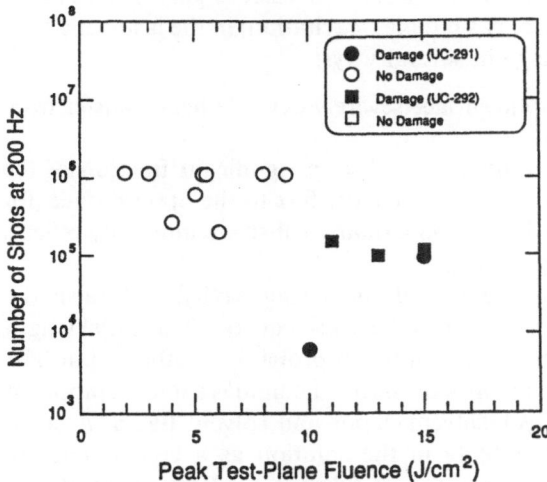

Fig. 4.7. Damage threshold measurement data at 308 nm for SiO_2 windows in contact with flowing dioxane

4.3.2 Particulates

Compared to gas flows, liquid flows are capable of transporting solids of similar density up to roughly 3 orders of magnitude greater in size [4.25]. The likely source of particulates in a system include the following:

1) System "dirt" (filter material, solvent impurities, dye impurities, etc.);
2) Dye or solvent degradation caused by the absorption of UV light, subsequent decomposition, and formation of large molecular clusters, which also absorb in the UV;
3) The precipitation of solid dye from solution because of rapid changes in local flow pressure or temperature. The system must be opened very carefully for draining or repair. Because the solvent is volatile, dye-particle deposition can occur on the windows and in the plumbing. Since the solubility is limited and often requires heat, the particulates left on the window do not necessarily go back into solution when the system is filled.

The most probable failure mechanisms involve the transport of foreign material (solid or gas) by the fluid to the optic surfaces. Particles near the wall enter the boundary layer, decelerate in the low-speed fluid, and come to rest on the surface. If the particle strongly absorbs in the UV, the subsequent energy transfer to the solid produces a high local heat flux, which the surface cannot immediately dissipate. This, in turn, produces local "chipping," which focuses subsequent pulse-beam energy and ultimately leads to catastrophic failure.

4.3.3 Bubble Formation

Because of sudden changes in flow conditions, local heating and lower surface energy at the particle can cause the generation, or release, of gas bubbles in the flow (cavitation). The formation of gas bubbles (cavitation) in the lasing medium may contribute to surface damage in several ways:

1) Bubbles that attach to the optic surface severely degrade heat transfer from the optics to the lasing fluid.
2) Bubbles on the optic surface internally reflect a significant fraction of the incident energy and increase the radiant energy flux to the optic surface (for example, a hemispherical air bubble on a quartz substrate internally reflects nearly 55% of the incoming energy).
3) Bubbles degrade the optical quality of the lasing cavity and raise the possibility of self-focusing of the pump and dye laser energy. The initial length scale of the bubbles released from the solution is probably less than 1 µm. The generation of bubbles is possible since the lasing medium is a four-component solution of gas (air), solid (dye), solvent vapor and solvent liquid. A small concentration of air is undoubtedly in the solution as a consequence of normal operation. The curves shown in Fig. 4.8 are intended to illustrate the solution's "bubble limit" (that point at which the gas first comes out of solution).

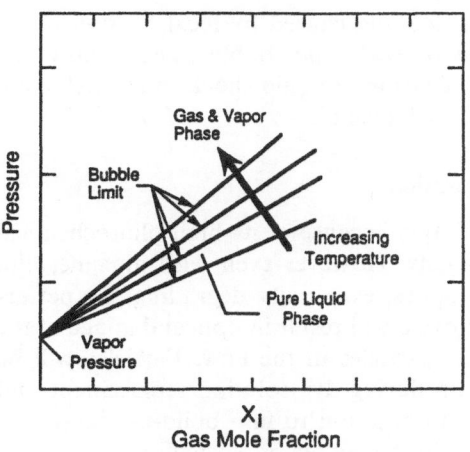

Fig. 4.8. Solute mole fraction as a function of pressure and temperature

For most dilute solutions, the isotherms are essentially straight lines with a slope proportional to P/X_i, where P is the system pressure and X_i is the gas solute mole fraction. The variation of slope with temperature is a characteristic of the gas–solvent system, and it is reasonable to assume that the air–solvent system exhibits this behavior. It follows that gas bubbles form when the temperature is increased at fixed pressure or when the pressure is decreased at fixed temperature. The bubble limit is sensitive to variations in temperature and pressure (Fig. 4.8). Gas bubble formation is a distinct possibility because significant pressure and temperature variations occur as the fluid travels through the mechanical pump and reservoir system and because the fluid accelerates and decelerates as changes in area that are associated with the various system components occur.

If a solid particulate comes to rest on the optic surfaces, another possible mechanism for local surface failure exists. Assuming that the particle absorbs strongly in the UV, the solvent in the local vicinity may receive sufficient energy to evaporate, creating a small vapor phase bubble. By itself, the formation of this bubble would produce a strong pressure wave (a shock wave) caused solely by the sudden change in local volume. However, the initial shock would be immediately followed by a rarefaction wave which rapidly reduces the fluid pressure.

Conservative estimates of the speeds of such waves in a liquid are on the order of 10^5 cm/s, which means that, on the optical surface, the stress would undergo an extremely rapid change from compression to tension. Moreover, after its initial formation, the bubble would be immediately cooled by the surrounding liquid and reabsorbed. The collapse of the bubble would be extremely rapid and would be accompanied by both a large local chemical energy release (the heat of evaporation) and the formation of high-speed jets of liquid that penetrate the bubble volume. The velocities of these jets have been estimated to be on the order of 10^4 cm/s.

The combination of rapid thermal loading caused by local heating of the particle, subsequent collapse of the associated vapor bubble, and rapid cyclic changes in surface stress due to the generation of liquid shocks and rarefaction waves thus suggests another damage mechanism.

4.3.4 Solvent Chemistry Particle Generation

Dioxane is a versatile solvent for UV dyes because of its high photochemical stability and excellent dissolving capacity. However even with dioxane, film deposits accumulate on the dye-cell optics, eventually degrading the performance of the oscillator and, if not removed, will result in optical damage of the cell. The primary source of the film is particles in the flow. Particles will be introduced into the dye laser flow loop during dye solution replacement and during dye-laser flow-loop operation. In addition to film build-up, scattering losses can also vary widely depending on the degree of particulate removal.

4.3.5 Other Sources of Particulates

Particulates can be introduced from several sources. The number of particulates in the unopened solvent bottles varied widely from supplier to supplier. Burdick and Jackson solvent was consistently the cleanest and also listed a recommended maximum shelf life on each bottle. Another source of particles in the dye solution of the dye laser is the operation of the flow loop. At least downstream of the dye cell in the turbulent wake region, dioxane dissolved or eroded aluminum. As a result, suspended metal and metal oxide particles exist in the dye solution. In general, stainless steel is the preferred construction material. For seals copper conflat gaskets, ethylene propylene (EPDM) and KALREZ O-rings, and a special epoxy recommended by Master Bond, Inc., of Hackensack, New Jersey, have been used. EPDM was the most common O-ring material used; however, when wet with dioxane, EPDM grows in length and cannot be reused until the material drys.

The dye itself can also be a source of particulates. While draining the system, care must be exercised to ensure that the dye stays in solution. Dye photoproducts and reaction products can be released. However, when a dye-dioxane flow loop was accidently allowed to warm to 50°C, a massive amount of contamination occurred from material being released from the filters and flow duct walls. Temperature control is essential for most dye solvent systems, and allowing a system to reach high temperatures will promote chemical reactions. Allowing the temperature to drop may result in the dye precipitating out of solution. We have observed dye precipitating on the cooler surface of the heat exchangers.

4.3.6 Filters

The solvent-dye solution should be filtered totally with each pass around the flow loop. The tightest, large-surface-area filters available today are 0.1 µm

Teflon filters available from either the PALL Corporation or the Millipore Corporation. PALL filters were used in the Los Alamos dye laser systems.

The use of the 0.1 μm filters greatly lengthened the dye-cell optics lifetime. Selection of the proper filters is very important because very small particles of dye and solvent reaction and decomposition products need to be removed to minimize optical damage of the dye cell caused by the buildup of reaction products. Coarser filter cartridges are not as effective at removing these particles. Selection of the proper filter material is also important to avoid reaction of the solvent and dye with the filter material and the production of particles from the filter material itself. For example, cotton filters produce particulates.

4.4 Fluid Issues

4.4.1 Flow Dynamics

Flow conditions can contribute significantly to the observed damage thresholds. The manner in which foreign matter is transported to the optical surfaces and the generation of local conditions that accelerate surface degradation are very important. The general fluid state and its transport and chemical properties are related to potential damage mechanisms in the text that follows.

Important fluid mechanical features of the cavity flow are shown in Fig. 4.9. The motion of the lasing fluid through the laser cavity serves several important functions:

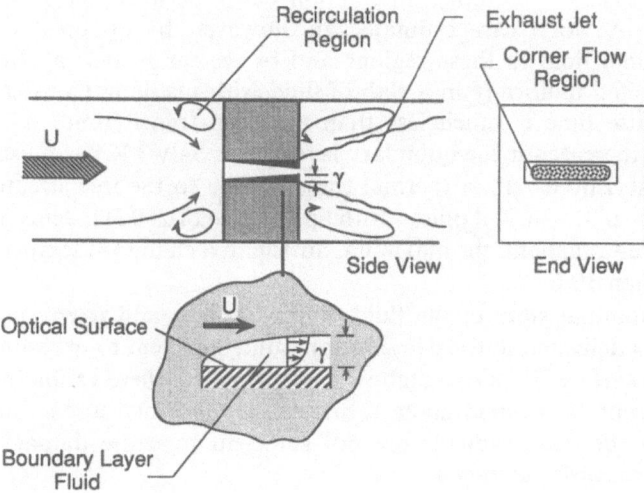

Fig. 4.9. General boundary and recirculation regions for an abrupt flow passage configuration

1) Unspent, optically active dye is presented to the pump beam at the beginning of each pulse.
2) The optical surfaces are cooled by convective heat transfer from the optical surface to the fluid.

To a first approximation, the solvent with dye may be considered an ideal incompressible solution. This assumption allows the analysis of fluid pressure, temperature and density variations to be physically uncoupled. In fact, pressure waves are produced by the deposition of energy but are very weak and travel with speeds significantly greater than the bulk fluid velocity (10^5 cm/s compared to 10^3 cm/s). Under normal conditions, these waves are quickly dissipated and do not produce significant fluid motion or density changes in the optical medium. For a solvent such as cyclohexane, the instantaneous temperature rise from the pump laser, assuming a uniform energy deposition, is less than 0.1 K for a case where the repetition rate is 250 Hz, the flow velocity is 1000 cm/s and the cavity fluid is exchanged approximately twice between pulses. Bulk heating of the cavity fluid is therefore unlikely to promote surface damage. The regions of high shear (boundary layers) near the liquid–solid surfaces shown in Fig. 4.9 may be characterized as either laminar or turbulent and are important features of the flow field. In these regions, the average fluid velocity varies from zero at the wall (no slip condition) to the free stream velocity at some distance d^*. Since the average speed of the fluid in the vicinity of the wall is low, both energy deposition and convective cooling of the surface are strongly influenced by the properties of this layer. Assuming uniform flow at locations far from the corner regions and assuming $d^* = 0$ at the cavity entrance, the mass displacement thickness, d^*, at the cavity exit is estimated to be 0.1 mm for either laminar or turbulent layers. Estimates of heat transfer in the vicinity of the cavity corners were not attempted because a complex interaction exists between the flows along adjacent walls. A worst case estimate can, however, be obtained by neglecting the additional fluid in these regions and by assuming that all the pump energy is deposited uniformly in a slab of fluid with maximum vertical dimension d^* (the pulse time is much less than the fluid transit time). The resulting temperature increase for the boundary layer gas is only 3 K per pulse. Using d^* as a characteristic length, a thermal transfer time to the free stream may be estimated to be 0.01 s or 2–3 pulses at the pulse rate of 250 Hz. Thus, if the fluid were completely stagnant, the maximum anticipated change in temperature would be less than 10 K.

Considering the nominal state of the fluid, flow velocities and cavity geometry, the energy flux delivered to the dye solution is not sufficient to promote thermal failure of the surface. Experimental evidence supports these estimates since most failures occur in isolated spots rather than uniformly across the surface. Although the thermal gradients are not sufficient to cause damage, beam quality can be seriously degraded.

4.4.2 Flow Uniformity

As part of the Los Alamos high-power dye-laser development, an off-line facility was constructed to study the flow uniformity. This apparatus is shown in Figs. 4.10 and 4.11.

The fluid used for the study was water. Geometrically and dynamically similar dye cells were installed for study. The dye cells were partially made of plexiglass to allow for flow visualization. The flow visualization techniques used include air bubbles, dye injection, a helium neon laser probe beam and a frequency variable strobe light. The initial goals of the experimental effort were as follows:

1) To increase the entrance and dye laser cavity flow uniformity,
2) To reduce the physical extent of the low-speed corner flow regions and potential particle deposition sites,
3) To minimize the potential for cavitation, and
4) To optimize the time required to clear the cavity of "spent" lasing fluid.

The recirculation zones (Fig. 4.9) on the inlet side of the dye cell extended throughout the inlet section, which is the entire top plexiglass section (the flow direction was from top to bottom). The region downstream of the dye cell was highly turbulent without well-developed circulation regions.

Fig. 4.10. Off-line flow facility for flow stability studies

Fig. 4.11. Flow cell mock-up for flow visualization

A variety of inlet and exit flow duct configurations were investigated. The flow-ducting geometries did not appreciably alter the inlet recirculation flow or the flow downstream of the dye cell. Several different screens were tried just above the top plexiglass region, and uniform potential flow was established using a 60-mesh screen throughout the top plexiglass region. The recirculation regions were barely observable, but the region downstream was still highly turbulent. The flow was found to be very uniform in the cavity region when probed with a helium neon laser. Hence, the region downstream did not appear to influence the flow throughout the dye cell. Based on this study, the dye oscillator and two dye-amplifier stages were modified to allow use of the inlet 60-mesh screen to establish a uniform dye cell inlet flow.

4.4.3 Dye-Cell Design Considerations

The initial Los Alamos dye cell placed the complexity in the fabrication, which resulted in excessive labor during assembly. Figure 4.12 shows a photograph of

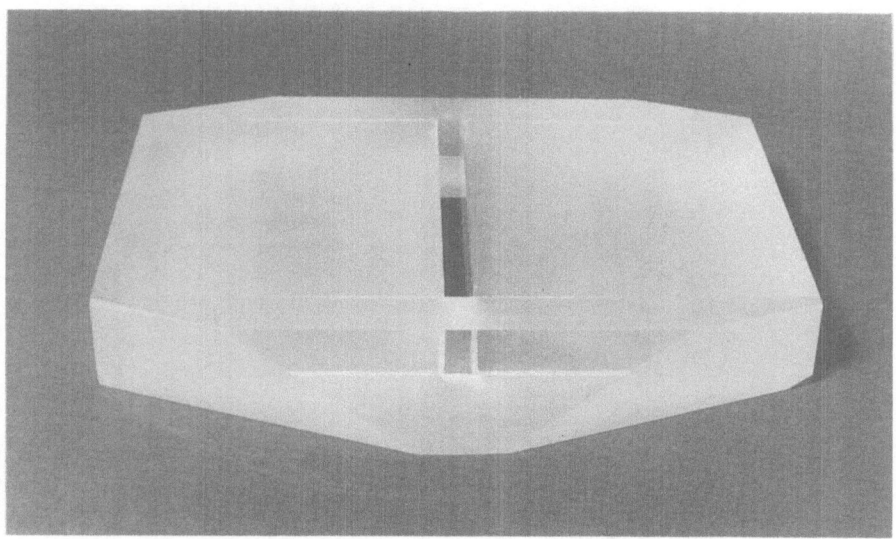

Fig. 4.12. Photograph of the quartz dye cell

the dye cell. The cell was made up of four pieces of precisely cut and polished quartz, glued together by an epoxy recommended by Master Bond, Inc., of Hackensack, New Jersey. Since dioxane is very nearly a universal solvent, finding a glue that would not be dissolved was difficult.

The change to dioxane and the subsequent increase in damage threshold allowed higher pump beam fluence and thus smaller pump beam area. The pump beam was reduced to 1.6 mm in height. A new design was developed to allow the use of standard round flat optical windows [4.3]. A photograph of the exterior of this new design is shown in Fig. 4.13.

In this design, the complexity is placed in the metal components, not in those made of quartz. The excimer laser pump windows in this design are 5 cm diameter quartz substrates, and the dye laser windows are 2.5 cm diameter quartz substrates. The flow direction is from top to bottom. A 60-mesh screen is used as a flow straightener between the top piece and the flow inlet section, which is a computer-designed contoured nozzle. The pump region flow cross section is 4 mm wide and 75 mm long. Downstream of the pump region is a constant angle diffuser section. A glue joint is still required on the 2 in. optics; however, the precision is now in the invar optic holders and not in the quartz. The most difficult part of the fabrication of this new cell is the electron-beam welding of the four pieces of the nozzle assembly together.

A cross-sectional view of this design is shown in Fig. 4.14, and the flow system for this dye cell is shown as a schematic in Fig. 4.15.

Fig. 4.13. Photograph of new flow cell design using standard optical elements

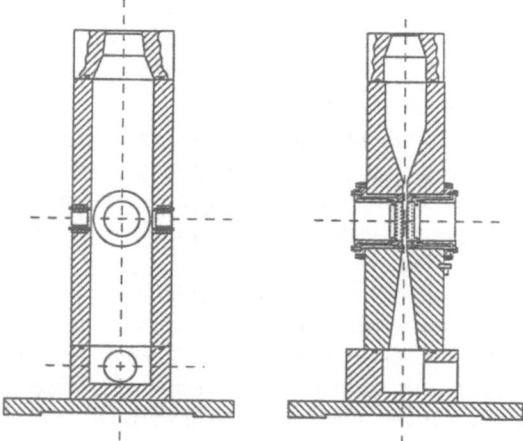

Fig. 4.14. Cross section views of new flow cell design

4.4.4 Flow Loop Pumps

The components, flow rates, and general operating conditions vary with choice of solvent, optical damage thresholds and number of dye laser stages. The two liquid pump designs shown below have several desirable features for systems

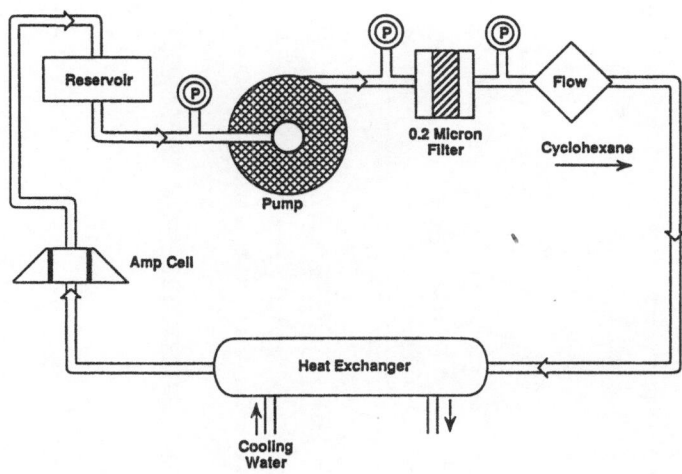

Fig. 4.15. Flow system schematic for the power oscillator

discussed above, which use solvents such as dioxane or cyclohexane. Table 4.5 summarizes the characteristics of two of these pumps.

Smaller and larger versions with the same general design features are also available. The Durco pumps have a much larger head capability and are much more forgiving on the design and pressure losses of the flow loop. However, in

Table 4.5. Characteristics of two suitable pumps for dye flow loops

	Durco Mark II Model 1–1/2 × 1–8	March Model TE-8K-MD
(water) Rated capacity	5.7 l/s, 7.6 atm	5 l/s at 2 atm
Motor	5.6 kW	3.7 kW
Seal	Mechanical rubbing face seal (silicon carbide)	Barrier with magnetic feedthrough
Drive	Direct with rubber	Magnetic coupling
Bearings	External–oil- lubricated ball bearings	Internal–ceramic on metal, fluid lubricated
Shaft	Stainless steel shaft drives impeller through mechanical seals	Ceramic–internal housing impeller rotates on stationary shaft
Housing	Stainless steel	Stainless steel
Impeller	Stainless steel	Stainless steel

Fig. 4.16. Compact portable flow system using the March pump

a carefully designed system, the March pumps can be smaller and much simpler in concept.

A compact, portable system is shown in Fig. 4.16. The essential components include a pump, a filter system, a surge tank/reservoir, a heat exchanger and a fill system. The instrumentation of this system consists of mechanical pressure gauges, a mechanical flow meter and thermocouples, interlocks and controls.

The most important component of the flow system is the flow pump, which must be matched to the pressure losses of the flow system. In this system the pump is a March Model No. TE-5.5S-MD, which pumps a maximum of 0.95 l/s with a system pressure drop of 570 mm Hg with dioxane solvent. The flow rate provided by this system was adequate to give a clearing ratio of 3 at 500 Hz for the final power oscillator. The motor power is 250 W. The seal is a barrier type with a magnetic coupling. The bearings are Rulon A. The shaft is alumina ceramic, the housing and impeller are stainless steel and the O-rings are EPDM.

4.5 Excimer Laser Considerations

4.5.1 Excimer Laser Pump

XeCl discharge lasers are now operating with lifetimes of up to 10^8 shots on one gas fill. These lifetimes can only be achieved through very careful engineering to avoid material contamination and, thus, are made with nickel-plated metals and ceramic construction only. Preionization with X-ray or corona gives no appreciable contamination; however, the spark preionizer is a major source of particulate contamination. Excimer lasers with X-ray or corona preionization generally have longer gas lifetimes than those using spark preionization. Window configurations of MgF_2 and CaF_2 are now available without the use of elastomeric materials. The pulse power circuits using ceramic capacitors and thyratrons and/or solid-state switches and several stages of magnetic pulse sharpeners have potential lifetimes exceeding 10^9 shots. Dust will be generated inside the laser because of the high temperatures at the surface of the electrodes and because of the presence of very active reaction products of the halogen. This dust may be collectable before it attaches to the window, but more work will be needed to achieve long-life windows without removing and cleaning them. Designs that provide gate valves at the windows so that the pressure vessel can be sealed off while the windows are replaced makes more sense for long-life service requirements. With this provision the laser gas can be retained in the laser, and exposure of the laser cavity to moisture can be avoided. The optimum repetition rate is still being debated. Time multiplexing with lasers operating at lower repetition rate has many appealing factors when considering the demands on the excimer design. The higher the repetition rate of the excimer, the higher the gas flow rates need to be and the more power must be expended to move this gas around and to remove the heat generated. The average power dissipated in all the loss mechanisms, as well as wear and erosion, goes up rapidly with increased repetition rate.

Multiplexing temporally with several pump lasers in a single dye cell may be better than increasing the repetition rate of the pump lasers. For the same reasons, the dye fluid flow rates must also be higher for high-repetition rates, and thus the losses, erosion, and chemistry are also more significant in high-repetition-rate dye lasers. Thus, time multiplexing the output of several lower-repetition-rate dye lasers may be preferred over increasing the repetition rate of a single-dye laser. Multiplexing by rotating mirrors or electro-optical scanning mirror systems may result in increased system reliability and lifetime over high-repetition-rate excimers and flow systems.

4.5.2 Excimer Laser Component Lifetime

Of most concern to the user are failure modes and estimates of the lifetimes of critical components. We shall address these concerns, and the expected down time, for major maintenance and routine gas changes.

4.5.3 Electrode Wear

The electrodes in discharge lasers erode with a resultant change in the electrode profile. A very hot corrosive plasma develops at the electrode surface, and no metal is immune to reacting in this environment. Of all the metals tried, nickel seems to be the most durable for the excimer laser electrodes. Less and more uniform erosion results when a very uniform discharge plasma is formed. For well-designed lasers, the discharge is free of arcing and, thus, the electrodes are not seriously pitted. If the preionization is not uniform and the electrode profile is not carefully shaped, the discharge will not be uniform and arcing can be observed. Gas contamination can also result in arcing, which will shorten the useful life of the electrode. With arcing the electrode will develop high points, which will increase the local electric field and will promote early breakdown and continued arcing. If care is exercised in developing the initial shape, spacing, uniform preionization, proper gas mixture of the electrode and in making it free of contaminants, the resultant excimer laser beam will be uniform and the electrodes will have long service time.

Preionization can be accomplished with a spark array, corona discharge, or X-rays [4.26, 27]. The spark preionizer is simple but typically erodes away faster than the discharge electrodes. It is difficult to produce uniform preionization using spark arrays, and if the spark preionizers are located to the side of the discharge electrodes, it is even more difficult to produce a sufficiently uniform discharge. In some of the popular laser designs, a set of peaking capacitors are charged through the preionization pins. The advantage of this design is the low inductance in this assembly and the rapid delivery of current to the discharge electrodes. However, the necessary high current density through the preionization pins limits the electrode lifetime. Erosion rates are typically 50 µg per coulomb. Limiting the coulomb transfer will extend the electrode life, and designs that use a separate circuit for the preionization and limit the current will exhibit longer lifetimes. Lifetime of the preionization electrodes can be more, typically 10^7 shots, for designs with high current density in the preionization pins.

Preionization with corona or X-rays results in a more uniform discharge and excimer laser beam. The electrode lifetime and the gas lifetime are usually longer with these preionization methods. However, achieving a high level of ionization is more difficult with these methods than with the spark array. Typically the ionization from corona or X-rays is 10^6 e/cm^3. Marginal levels of ionization can also lead to arcing. The preionization electrodes will exhibit long life for these methods, but the discharge electrodes may not exhibit long life if the ionization level is marginal.

In some designs the corona or X-ray preionization is directed to the discharge gap through one of the electrodes. This location is ideal for the ionization to uniformly illuminate the discharge region, compared with preionization located off the side. However, the transparent electrode must be very thin or perforated to transmit the ionizing X-rays or UV. This electrode is much less robust and typically will deteriorate faster than the solid opposite electrode. Foil electrodes may not have lifetimes much over 10^7 shots. Solid electrodes plated with Ni do not have as long a lifetime as solid electrodes, because the Ni plating tends to flake and peel after the plating is worn down.

4.5.4 Window Lifetime

The best choice of window materials for the excimer lasers is MgF or CaF_2, and MgF is superior to CaF_2 in strength and increased thermal conductivity. However, both are much preferred to quartz as a window material. All these window materials will exhibit some chemical attack; however, quartz is particularly vulnerable. For high repetition rate and thus higher average power deposited in the window, MgF is the best choice.

The mechanisms for window failure are as follows: The very hot reactive plasma reacts with the laser electrodes, and the halogen combines with the laser electrodes to produce a large quantity of metal chlorides for the XeCl laser and fluorides in the case of XeF and KrF. These particles are deposited throughout the laser cavity and will also coat the windows. Purging the windows with laser gas can be helpful, but most laser builders do not offer this option nor do they adequately keep the particles off the windows. These particles are very absorptive and tend to burn on the window. Eventually the window has excessive absorption with the resultant loss in laser output and increased heating of the window. At high repetition rate, the deposited energy in the window can be sufficient to break it.

Organic materials in the laser cavity, such as O-rings, dielectric materials for the laser insulators and lubricants used for bearings, will react with the halogen and radicals produced in the discharge and will also deposit on the windows. We have found that these products can etch even MgF windows. Typically, windows need cleaning every 10^6 shots, even with electrostatic precipitator dust collection. If extra care is used to purge the windows with clean dust-free gas and if organic materials are avoided in the laser cavity, the window lifetimes can be extended to 10^7 shots. Using gate valves at the windows so that the laser cavity can be closed and the windows removed without exposing the laser cavity to air will allow rapid window removal, cleaning, and reinstallation without the need to refill the laser.

4.5.5 Bearing Lifetime

The laser gas must be circulated to remove the hot ions from the discharge region between pulses. The clearing ratio needed to establish an arc-free discharge has to be at least 2 and typically is about 4–6. Most commercially available lasers use an internal fan assembly that is bearing supported. The lubricant used is a halocarbon grease such as Krytox. Depending on the care exercised in keeping the laser free of moisture, the bearing life can be as long as the electrode life. It is good practice when changing electrodes to change bearings.

4.5.6 Heat Exchanger Lifetime

Heat generated by the discharge and by the fan must be removed. Most excimer lasers use a water-to-gas heat exchanger in the gas flow duct. Typically, the heat exchanger tubing and fins are aluminum. The most severe failure of these exchangers is caused by water leaks inside the pressure vessel. Water leaks in the laser cavity of course create a major contaminant; lasing ceases and corrosion begins. Leaks are most likely at tube joints. In our studies, we found another major cause of leaks. The cooling water at the Los Alamos facility has a pH of 8.5, which is very basic, and may be typical of many industrial facilities. Aluminum will corrode rapidly if the pH is not near 7 (neutral). To overcome this problem, we used stainless heat exchangers; however, care must be exercised in making connections to avoid electrolytic corrosion enhancement effects. We used a separate cooling loop on our excimer laser systems so that the pH could be controlled and antifreeze, which has pH buffering agents to prevent corrosion, could be added.

4.5.7 Gas Lifetime

The excimer laser performance is critically dependent upon the laser gas quality. The most likely cause of output or beam quality loss is gas condition. Two basic problems can develop: contamination or halogen depletion. We will consider most of the sources of contamination, ways of minimizing the contaminants and how to trap those that are generated. We will also discuss the mechanisms of halogen depletion and various methods of correction.

4.5.8 Excimer Laser Gas Contamination Sources

Gas Supply. As-received gas can contain contaminants that can be deleterious to gas lifetime and can affect the discharge-pumping kinetics and the production of absorbers. High-purity gases should be used with low amounts of H_2O and CO_2.

Gas Cylinders. The cylinders used for gas storage must be well passivated and free of residual contaminants.

Transfer Lines. The gas transfer lines must be evacuated and purged before filling. Plastic lines should not be used. Leaks in the system will contaminate the gas and destroy the gas lifetime. We found laser systems that have been He-leak checked and can be pumped down to 10^{-7} Torr provide the best performance.

Exposure to Air. Any time the system is opened the chance for contamination is high. Opening the laser, even for quick window replacement, can be an opportunity for contamination. It is best to purge the laser with He or Ar while the window replacement is being done. A small outward flow of inert gas can help. After replacement of the windows, the laser should be evacuated and reconditioned by discharging in laser gas mix.

Vacuum Pump Oil. Vacuum pumps, if not properly trapped, can introduce oil by back streaming into the laser cavity. Oil introduced in the laser cavity will become a carbon deposit on the windows and will degrade the laser gas lifetime.

Extended Exposure. If the laser is opened for an extended period, there is a high risk of corrosion. The metal chloride dust that collects in the laser with long runs is highly hygroscopic. Laser parts with chloride coatings will become noticeably wet within a few hours after exposure to air. This solution is very corrosive and if not removed can damage the components. Even the dielectric materials can exhibit corrosion damage.

Cleaning and Passivation. For most of today's commercially available excimer lasers, it appears that the best cleaning and conditioning approach is to take the laser apart periodically and wash out the particulate residues with soap and water followed by a water flush. When reassembled, the laser should be helium-leak checked to the 10^{-8} scc/s level. The laser should then be conditioned by running it for several days to several weeks with a halogen-containing gas. Inadequate passivation conditioning will result in limited gas lifetime and poor laser performance.

Laser Component Materials. Excimer lasers constructed of metals and ceramics with little or no organic materials exposed to the halogen and discharge products will exhibit much longer gas lifetimes than those constructed with organic materials. In an extensive study of laser material chemistry [4.28–31], *Tennant* found that organic materials such as viton O-rings and kynar insulation materials react with the halogen. For near-room temperature operation from regions impacted by the discharge or preionizers, stainless steel, nickel or electroless nickel-coated aluminum are adequate as a general construction material. Stainless steel works very well as long as the system is dry; however, if not properly cleaned after HCl exposure, subsequent exposure to moist air can lead to severe corrosion problems upon exposure to air, but inadequate cleaning and preparation of the material can lead to substantial depletion of the HCl and to the production of life-limiting reaction products. In general, the use of plastics can and should be avoided because of problems with HCl depletion and gaseous contaminant formation. Once the laser system is clean and dry, HCl stability, even low concentrations such as those found in a XeCl laser, could potentially give a shelf life of many months to a year or more.

4.5.9 Excimer Laser Gas Halogen Depletion

Typical gas mixes for the XeCl laser have about 1–3 Torr partial pressure of HCl, 20 Torr Xe balance Ne or He. The major change in gas composition during run time is loss of the halogen caused by the chemical reactions discussed above. The hot spark in the spark preionizers contributes to a more rapid depletion of the halogen than that occurring in the main discharge. However if arcing is observed in the main discharge, this too can rapidly burn up the halogen. Depletion of the halogen can be compensated by introducing a small amount of HCl when the laser output begins to drop. The typical lifetime of a single gas fill for a well-passivated and leak-tight laser is 10^6 shots, which can be extended by a factor of 2 or so by introducing HCl. If a cryogenic gas clean-up system is used to trap contaminants and reaction products, gas lifetimes of over 10^7 shots can be expected. To obtain long optic and gas lifetimes with today's lasers, it is necessary to use a gas processor and a halogen concentration control process.

4.6 Gas Processors

If the gas in the laser builds up significant reaction products, it either needs to be reprocessed or dumped. Xenon currently costs about $15.00 per standard liter. Hence, applications requiring long-term, day-after-day running can quickly result in the gas lifetime being a dominant issue. The predominate contaminants appear to be H_2, CO_2 and other carbon-containing compounds. Only CO_2 and chlorocarbons need to be removed, which could be done by pumping the gas through a cryogenic trap operating at about $-143°C$. If H_2 must be removed, then a high-temperature getter trap or the addition of Cl_2 to a cryogenic system would have to be used. The two most applicable types of gas processors are getter traps and cryogenic traps, and the most versatile is the getter-trap-type system. Getter-trap systems operate at $650°C$ thus, for safety reasons, were not seriously considered for the Los Alamos laser system because of the presence of large quantities of highly flammable solvents. Instead, the cryogenic approach was investigated, and we discovered that, for a large XMR XeCl laser, it was theoretically possible to run the cryogenic trap at a temperature cold enough to remove most of the expected carbon-containing gaseous contaminants and to control the HCl concentration (by charging the trap with excess HCl) at the desired level without changing the Xe concentration. Lambda Physik tested one of their lasers at a slightly higher temperature and achieved a major improvement in the window lifetime. The Lambda Physik laser operates at a higher Xe partial pressure than the XMR laser, and hence, a higher operating temperature is necessary to avoid removing the Xe. Unfortunately, the higher operating temperature eliminates the option of using the cryogenic trap to control the HCl partial pressure on the Lambda Physik laser. An added feature of the cryogenic system is that it can also be charged with Cl_2 to automatically control the H_2 buildup problem, which may occur during long-term operation of XeCl lasers. The Cl_2 partial pressure would be low enough so that the absorption at 308 nm would be insignificant but high enough to remove the H_2 as it accumulates.

A great deal of progress has been made in the last 11 years in understanding HCl material-interaction problems in XeCl discharge lasers. R. Tennant made extensive studies of excimer laser gas lifetime and the effects of various materials and contaminants. Any laser being selected for long-life service should be reviewed in light of his studies on materials and contamination. *Tennant* [4.28] found in 1981 that, for well-conditioned XeCl lasers, several million shots were possible before the laser output drops significantly, with the depletion of the HCl being the major reason. He presented results showing that over 8000 discharges per discharge volume can be accumulated for about 20% output drop and, with halogen injection, can recover to full output. The discharge volume he used is the electrode length times the electrode separation times the diameter of the laser beam. This volume represents the approximate volume of gas chemically altered by a laser electric discharge. Lasers with a large reservoir of gas relative to the discharge volume will exhibit longer run times before loss of power. Lifetime estimates can be made for a particular laser by knowing the volume of gas in the laser divided by the discharge volume times 8000. Lasers that have been carefully constructed and are free of organics will exhibit even much longer gas lifetimes.

4.6.1 Typical Excimer Laser Gas Lifetime Performance

Jursich et al. [4.32] recently surveyed the status of commercial excimer lasers for gas lifetime performance. Cymer Laser Technologies gives a performance specification of 8 h/fill at a maximum repetition rate of 200 Hz (5.76×10^6 shots) for the 50% power point. This data is for operation with KrF, and although not reported, a much longer life would be expected with this laser operating on XeCl. Lambda Physik's LPX 220i KrF lasers have demonstrated 10 h run times (7.2×10^6 shots) before halogen injection to restore power. Lumonics reports their EX-600 demonstrated 75×10^6 shots on a single ArF gas fill using a gas processor. Their 100 W KrF laser operated for 6×10^6 shots without a gas processor. XMR demonstrated their XMR 150 industrial XeCl laser to have 250 h of continuous full-power operation at 300 Hz with one intermediate gas change. They pretest all their lasers at full power for 15×10^6 shots on a single gas fill without a gas processing or halogen injection. This XeCl laser exposes only nickel and Teflon to the laser gas. Also this laser uses corona preionization. Choosing these materials and ionization, combined with a rather large reservoir of laser gas, gives this laser superior gas lifetimes. The Lambda Physik EMG 203 MSC laser used for the Los Alamos high-power dye laser would typically exceed 6×10^6 shots per gas fill before the power dropped by 30%. In one test on gas lifetime at 250 Hz (62.5 W operation) with new gas fill and clean windows, the laser ran 22 h (1.8×10^7 shots) before the power dropped to 50% of the initial power. At least 20% of that power loss was due to window clouding caused by chloride dust buildup. Tennant suggests that, if a few basic clean-up procedures are followed and proper material selection is made, XeCl lasers can have gas lifetimes reaching 10^{10} shots. *Hans* and *Scott* [4.33] give a ten-point guideline for preventing contamination that will significantly aid in achieving long gas lifetimes.

4.6.2 Beam Characteristics

a) Temporal

The typical temporal profile of XeCl lasers is given in Fig. 4.17. The FWHM is between 10 and 30 ns for standard oscillator configurations. Since dye lasers require high-pumping-power densities for efficient performance, it is desirable that the excimer pulse be short and not exhibit a long tail. Very little of the energy in a long-tail-type pulse will be of any value in pumping a dye, because the power density in the tail is insufficient to pump the dye efficiently. The laser gas mix will affect the lasing pulse temporal profile. Using Ne instead of He will increase the pulse width. The pulse width also decreases as the laser gas deteriorates at long run times. It is not only important to determine that the temporal pulse profile is sufficient initially but also to monitor the pulse with time to verify that pumping has not degraded.

b) Spatial Intensity Profile

The intensity variation across the laser beam is typically very flat when measured from electrode to electrode. The intensity profile is more Gaussian when measured at 90 degrees to that axis, with a peak-to-average value near 2. This value affects the gain distribution in the dye cavity and also limits the pump fluence since the damage threshold is defined by the peak fluence. The use of beam homogenizers are recommended to improve these limitations and are discussed elsewhere in this chapter.

c) Beam Quality

The excimer lasers are very high gain and thus exhibit an appreciable amount of ASE. The beam divergence for standard resonator optics is typically 2–6 mrad. With the use of unstable resonator optics and aperturing to restrict the transverse modes, the divergence can be an order of magnitude smaller. For pumping a dye laser the requirements for low divergence are minimal, and with proper relay optics, the dye cell can be adequately pumped with standard resonator optics on the excimer.

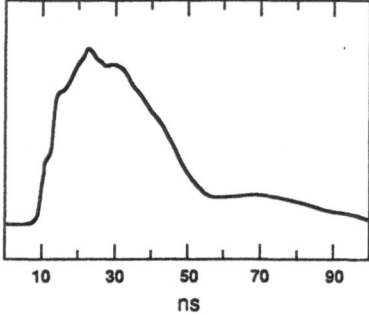

Fig. 4.17. Temporal shape of the EMG-203 MSC excimer laser pump pulse

4.6.3 Jitter

Shot-to-shot timing errors can occur because of triggering problems with the high-power switch. Most high-power excimer lasers today use thyratron switching. If the thyratron is properly triggered, the jitter can be no more than 2 ns; however, as much as a 10 ns jitter can occur if care is not exercised. MOPA designs require that the amplifier be pumped and the gain medium established when the seed pulse arrives from the oscillator. Timing errors of 1 ns can make a significant difference in the dye laser output. Similarly, timing errors of 1 ns between two excimer pump lasers that pump a single cell can produce significant output differences. Excimer lasers using spark gap switches are even more likely to have jitter and timing stability problems. Magnetic pulse-sharpened lasers also exhibit jitter caused by the magnetic circuit. The magnetic material switches from a high-permeability state to a low-permeability state when it reaches magnetic saturation. The saturation value is a certain number of volt seconds. If the charging voltage is low, the time to saturation is lengthened to reach the fixed volt-second saturation. Voltage regulation must be extremely accurate to keep the time to saturation within 1 ns. Further, the saturation value is a function of temperature, so the timing can drift if the magnetic material temperature changes.

4.6.4 Timing Drift

As mentioned, drift (slow changes in timing) can occur because of temperature change in the magnetic circuit. Timing drift also occurs when the thyratrons heat up, and we discovered that changing the laser repetition rate from 125 Hz to 200 Hz results in additional heating of the thyratron, causing the thyratron to change its timing characteristics. Thermal-induced timing drift is especially obvious at start up of the laser. Typically it takes about 1 hr for temperatures to reach equilibrium. Initial timing drift can be as much as 30 ns. Temperature-induced timing errors of course occur over minutes, and manual adjustment can usually be done to correct this error. In the Los Alamos dye laser facility, we monitored the thyratron switching by placing a CVR (current viewing resistor) in the cathode return path to ground. This signal was used to detect when the laser fired and was compared with the trigger demand. The error in the timing was then averaged and used as a correction signal to a digital delay generator so that the timing trigger pulse was changed to correct for drift. The automatic control system shown below controlled six excimer lasers and kept the error within 1 ns even through the warm-up period.

4.6.5 Automatic Timing Control Loop

Figure 4.18 illustrates this automatic system. The CVR had a resistance of 0.001 Ohms and gave a signal of about 6 V when the thyratron fired. Although the risetime was fairly slow, it served as a reliable signal to indicate when the

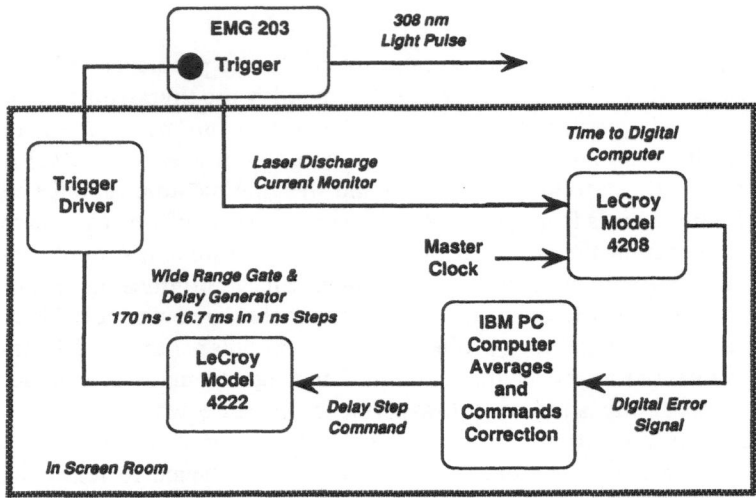

Fig. 4.18. Dye laser timing system block diagram

laser fired. Probably the optimum signal would be from an optical pickoff; however, the CVR signal is robust and independent of optical alignment. The CVR signal was then sent to the LeCroy digitizer, which compared the time arrival of this signal with the reference trigger signal. This timing difference was then digitized and fed to the IBM XT computer. An average of 10 of these differences were then compared to the demand time entered in the computer. The difference in the demand time and the average of the measured values was then the error signal, which was sent by the computer to a LeCroy delay generator to shift the time of the trigger sent to the laser. So we corrected the trigger input to the laser to correct for drift. The computer would display a table where the desired delay could be entered and then would continuously display the error signals, showing how well the system was keeping up with the drift. The drift was kept well within 2 ns for hours of run time on all lasers.

4.7 Dye Laser

4.7.1 Specific Design

Reviewing the design details of the Los Alamos high-power dye laser and its performance characteristics provides insight into the problems and the success of UV dye lasers that produce over 150 W of both MOPA and power oscillator design.

The original design was conservative in estimating the dye efficiency in achieving the desired UV dye laser energy and beam quality. For high beam quality, a MOPA system was designed that used a small, well-apertured oscillator to establish a single longitudinal mode. The MOPA system also incorporated brewster angles to establish high polarization, to help reduce the problem of parasitic oscillations developing, and to minimize reflection losses. A precisely positioned prism was used to achieve wavelength control. The oscillator output was followed by a preamplifier. The beam was transported from the oscillator to the preamplifier by a telescope pair of lenses. Two additional amplifier stages were incorporated to bring the energy per pulse up to the required level. The original design was based on the following: The pump beam fluence was limited by the optical damage and the two-photon absorption in the dye. At 308 nm, high-quality UV quartz will damage at 8 J/cm^2 and lower grade quartz will damage at 2 J/cm^2. The original design assumed that a total of 500 mJ/pulse would be delivered to the large amplifier cell by the two Lambda Physik EMG 203 excimer lasers. This energy was to be spread over the pump window of 35×7 cm^2 or 2.54 cm^2, for a fluence of 204 mJ/cm^2. Since this threshold was an order of magnitude below the single-shot threshold for damage, it was considered appropriate for high-repetition-rate operation. The optical damage threshold depends on the pump intensity, the dynamics of the dye, the dye concentration and the dye laser intensity.

Experiments indicated that this threshold might be a problem at fluences as low as 100 mJ/cm^2. The minimum pump fluence was set considering the need to saturate the dye. Experimentally, the dye cell saturation was determined for a dye fluence of about 16 mJ/cm^2. Assuming an input energy of 2 mJ of dye beam into the first amplifier gives an amplifier area of 0.1 cm^2 (the dye volume is this area times the length). Flow considerations for clearing this volume between laser shots at a 500 Hz repetition rate without flow instabilities resulted in choosing a volume of $3.5 \times 3.5 \times 70$ mm^3. The maximum dye laser fluence is constrained by optical damage. The optical damage at 390–400 nm was assumed to be much less than that measured at 351 nm where good data existed. We estimated that quartz should have a damage threshold of about 8 J/cm^2. So the design was to keep the dye laser beam fluence at a design point below 2 J/cm^2. This indicated that the maximum dye-beam energy, propagated through the amplifier exit window, would be 245 mJ. Despite these plans the damage threshold appeared to be much worse than estimated, and damage was observed in both the pump windows and the dye-beam windows. This damage was rapid and would not allow operation for more than a few minutes.

The dye volume was only $1 \times 1 \times 24$ mm, and the dye flow was transverse to the excimer optical pump as shown. One EMG 203 excimer was used to pump both sides of the dye cavity with beam splitters. The original design provided a longer temporal pumping pulse to give the cavity more time to establish single-mode operation. This was accomplished by splitting some of the excimer beam and delaying this pulse to effectively stretch the pump pulse from an anticipated 20 ns to near 30 ns. The excimer beams were then shaped by spherical and cylindrical lenses to develop a long (24 mm $\times$ 1 mm high) pumping beam, which excited the dye transversely for the full length of the optical cavity.

The oscillator dye cell shown in Fig. 4.12 was fabricated from quartz pieces that were glued together and then repolished on the top and bottom so that a seal could be made with an O-ring. The cell also had 6° tilted internal dye laser beam windows to minimize internal parasitic oscillations. This construction produced a high-quality, low-divergence beam of about 3 or 4 mJ/pulse, depending on the size of the adjustable aperture. The excimer pump laser produced near 400 mJ, but the transport efficiency was low because of the great number of optical surfaces in the path to the dye-cell window. Despite the fact that the optics were AR coated, only about 80% of the pump beam arrived at the dye cell. Considerable work was done in optimizing the dye concentration, pump beam power, output coupler fraction and aperture. This initial design would have been successful if the dye-cell window damage thresholds had been near 2 J/cm^2 as anticipated from previous experiments on dry quartz. Severe window damage and black deposits occur rather rapidly on both the dye pump windows and the dye exit windows. The dye concentration was fairly high in this early dye-cell design because the dye-cell width was only 1 mm, and the excimer beam was designed to be nearly 80% absorbed over this short distance. The dye concentration was 150 mg/l. This level was more than the amount of TBS that could be dissolved and stay in solution with cyclohexane. Many factors led to the rapid deterioration of the windows, one of which was undissolved dye particles that adhered to the window. At this early stage not much was known about TBS. In working with concentrated dye solutions, we found that the cyclohexane would evaporate away leaving a film of dye particles on the wall of the vessel. In the early testing stages of this system, the dye cell was drained without realizing that the dye cell wall could be left with an absorbing film.

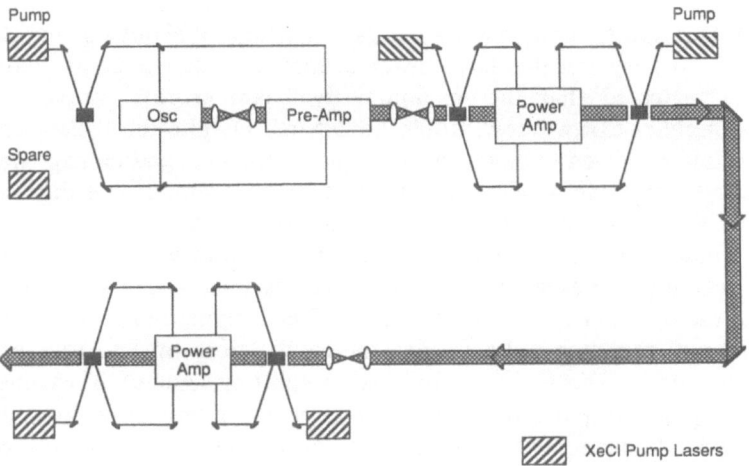

Fig. 4.19. Original MOPA dye laser system with three stages of amplification

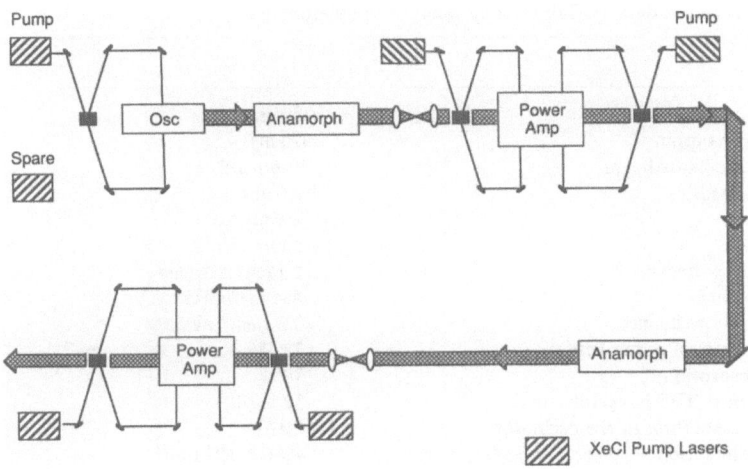

Fig. 4.20. Reconfigured MOPA using only two stages of amplification

The preamplifier was identical in cell geometry. In subsequent tests, the preamplifier was removed because the oscillator was able to produce sufficient energy to saturate the first large amplifier. Originally the preamplifier was pumped with the same excimer that pumped the oscillator by beam splitting some of the pump energy. Figure 4.19 shows this system.

Major reconfiguration was required to get this system to work reliably. The fluence of the pump and dye laser beams was decreased to the level required to eliminate damage. Improved optics for beam handling and beam shaping were introduced. Decreasing the fluence on the pump windows was accomplished by increasing the height of the beams, which caused the beams to be elongated in height. Accommodation of the required change in aspect ratio of the beams from one amplifier to the next was accomplished by use of anamorphic prism pairs. Pairs of lenses were incorporated as Fourier transform pairs to control the beam transport around the optical tables and to reimage the oscillator beam inside the amplifier-pumped volume. Also, the output of the first amplifier was reimaged in the second amplifier-pumped volume. Figure 4.20 shows the reconfigured optical system.

This scheme was far from the optimum pumping conditions to achieve high efficiency; however, it did result in essentially eliminating the optical damage problems. The required energy was produced so that experiments could proceed while the causes for the very low damage threshold were resolved.

4.7.2 Experimental Results with Cyclohexane Solvent

Table 4.6 shows the measured performance for He oscillator data, and Tables 4.7 and 4.8 give the performance data for the following two amplifier stages. The

Table 4.6. Performance data on TBS/cyclohexane dye oscillator

Oscillator	
Wavelength	395 nm
Oscillator energy output	20 mJ
Divergence of oscillator beam	5 mm mR
Dye-beam bandwidth	0.6 nm
Dye-beam width	4 mm ($1/e^2$)
Dye-beam height	4 mm ($1/e^2$)
Dye-beam window fluence	1 J/cm^2 average
Excimer pump energy	484 mJ/pulse
Pump beam window fluence	0.2 J/cm^2 average
Transport efficiency of pump beam	77.5%
Pump beam absorption	90%
Dye concentration: TBS in cyclohexane	32 mg/l
Flow rate of dye solution in the oscillator	0.27 l/s
Flow passage dimension	$4 \times 6.7 \times 27$ mm^3
Pumped volume	$4 \times 4 \times 25$ mm^3
Dye flow clearing ratio	2 at 125 Hz

Table 4.7. Performance data for TBS/cyclohexane dye amplifier I

Amplifier I		
Beam transport efficiency to amplifier I		77.4%
Input energy to amplifier I		15 mJ
Output energy of amplifier I	(minimum)	60 mJ
	(maximum)	110 mJ
Output window fluence	(average)	0.15 J/cm^2
	(maximum)	0.28 J/cm^2
Excimer pump energy at cell window I		320 mJ
Excimer pump energy at opposing cell window		236 mJ
Efficiency of pump beam transport		80%
Pump window fluence		0.1 J/cm^2
Anamorphic prism aspect ratio	(in)	4.2×3.5 mm^2
2.7 × vertical	(out)	4.2×9.4 mm^2
Beam size at amplifier I input		4.2×9.4 mm^2
Dye-cell flow passage		$5 \times 19 \times 70$ mm^3
Amplifier dye-cell-pumped volume		$5 \times 9.4 \times 65$ mm^3
Amplifier I dye solution flow rate[a]		2 l/s
Dye concentration in amplifier I		32 mg/l

[a] Both the oscillator cell and the amplifier I cell are in the same dye loop.
The oscillator flow was restricted to 0.3 l/s.

dye beam spatial profile for both horizontal and vertical directions is given in Fig. 4.21.

The spatial profile of the amplifier I output beam, both vertical and horizontal, is given in Fig. 4.22. The amplifier I saturation curve was measured and is shown in Fig. 4.23.

Table 4.8. Performance data for TBS/cyclohexane amplifier II

Amplifier II		
Beam transport efficiency to amplifier I to II		89%
Input energy in dye beam to amplifier II	(nominal)	53 mJ
	(maximum)	97 mJ
Input beam size into amplifier II		5×7.4 mm^2
Output beam energy	(nominal)	110 mJ
	(maximum)	160 mJ
Cell window output beam fluence		0.29 J/cm^2
Excimer pump beam energy at cell window		272 mJ
Excimer pump beam energy at opposite cell window		366 mJ
Amplifier II pump beam transport efficiency		80%
Amplifier II output vertical	$(1/e^2)$	7.2 mm
Amplifier II output horizontal width	$(1/e^2)$	5.2 mm
Beam divergence of amplifier II output	(vertical)	14.4 mm mR
	(horizontal)	9.4 mm mR
Dye-cell flow passage volume		$5 \times 19 \times 70$ mm^3
Dye solution flow rate		3.2 l/s
Dye concentration		25 mg/l
Pumped volume		$5 \times 7.4 \times 65$ mm^3

Dye Laser Oscillator

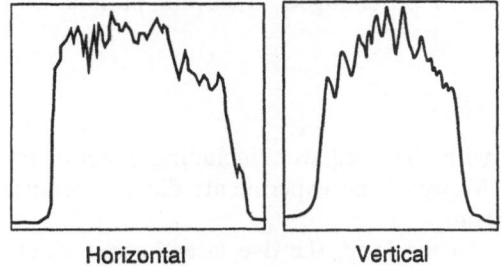

Horizontal   Vertical

Fig. 4.21. Typical beam profiles for the oscillator for TBS/cyclohexane dye laser

TBS/Cyclohexane Dye Laser

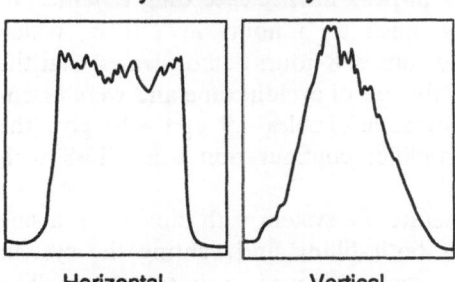

Horizontal   Vertical

Fig. 4.22. Typical beam profiles for the first amplifier stage for TBS/cyclohexane dye laser

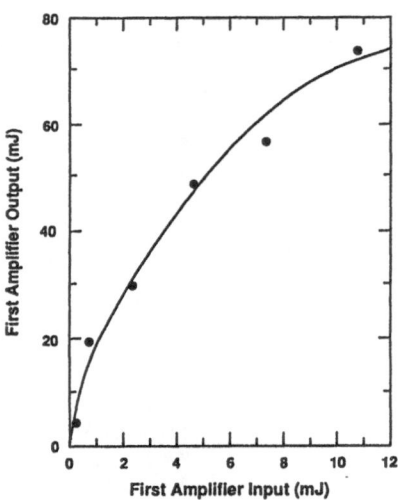

Fig. 4.23. Saturation curve for the first amplifier for the TBS/cyclohexane dye laser

Experiments were conducted with the dye laser system in this configuration until the causes of the low-damage threshold were determined. Early communication with Lambda Physik indicated stability problems with cyclohexane and low solubility of TBS. The decision to use cyclohexane instead of dioxane was a matter of safety rather than of choosing the optimum solvent. Dioxane was much more hazardous – it is tumorigenic and is absorbed through the skin. All the plumbing and O-ring seals had to be changed to be compatible with dioxane.

4.7.3 Experimental Results with Dioxane

Various configurations were tried with the oscillator, including experiments without the prism-tuning element. Many of the experiments did not require tuning away from line center of 400 nm.

By adding new dye concentrate to the loop, the dye laser output would recover. No difficulties with damaged dye molecules or fragments causing absorber species were observed. We believe that the products were being trapped in the filter system or on the plumbing walls. The best result was that there was no sign of any optical damage at peak fluence exceeding 1 J/cm². No window damage was observed for as much as 3 hours at 250 Hz, which amounted to 2.7×10^6 shots. Subsequent runs of 8 hours demonstrated that the damage problems were associated with the use of cyclohexane and were essentially eliminated with the change to dioxane. Tables 4.9 and 4.10 give the performance data for the oscillator-amplifier configuration using TBS with dioxane solvent

Considerable effort was made to operate the system with dioxane in a safe way. A new method was adopted for both filling and venting the system remotely. A very sensitive sniffer system was employed to detect the smallest amount of vapor, which was well below the explosion level.

Table 4.9. Performance data of the oscillator with TBS/dioxane dye

Wavelength		396 nm
Bandwidth	(FWHM)	2.3 nm
Oscillator output energy	(100 Hz)	7 mJ (max)
	(maximum)	80 mJ
Excimer pump input to the cell window		510 mJ
Excimer beam size at the cell window		3.4×23 mm^2
Excimer fluence at the cell window		0.65 J/cm^2
Efficiency of the dye oscillator		18.5%
Excimer fluence at the cell right window	(average)	0.25 J/cm^2
	(peak)	0.5 J/cm^2
Excimer fluence at the cell left window		0.23 J/cm^2
Dye laser fluence at the output window	(average)	0.5 J/cm^2
Dye lifetime in photons/molecule of dye[a]	(for half the output energy)	4×10^3

[a] Should be compared with quoted half energy values stated by Lambda for QUI at 50×10^3 photons/molecule.

Table 4.10. MOPA system performance with TBS/dioxane

Oscillator output		80 mJ
Oscillator wavelength		401.5 nm
Excimer pump energy to the dye cell		400 mJ
Efficiency of the dye laser		20%
(output energy/excimer input to the dye cell)		
Dye-cell pumped volume		$3.4 \times 4 \times 23$ mm^3
Dye solution flow rate		0.56 l/s
Dye concentration		30 mg/l

Amplifier I

Input beam size		4×6 mm^2
Input beam energy		40 mJ
Output beam energy	(average)	150 mJ
	(maximum)	200 mJ
Output beam spatial profile	(vertical)	15 mm mR
	(horizontal)	10 mm mR
Output window fluence	(average)	0.6 J/cm^2
	(peak fluence)	1.2 J/cm^2
Excimer laser input energy to cell windows		782 mJ
Excimer pump fluence at the cell window		0.1 J/cm^2
Dye cell flow passage volume		$5 \times 19 \times 70$ mm^3
Dye flow rate		3.8 l/s
Dye concentration		30 mg/l
Beam transport loss from oscillator to Amplifier I		80%

4.7.4 Dye Laser Configuration Using Dioxane Solvent

Figure 4.24 shows the laser configuration, Table 4.10 the measured performance values for this system, and Fig. 4.25 the saturation curve for amplifier I.

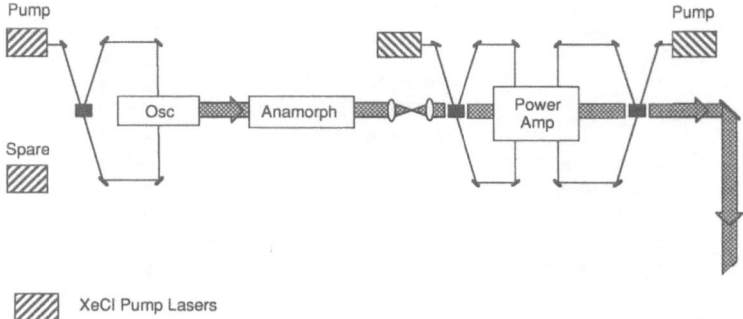

Fig. 4.24. MOPA configuration for the TBS/dioxane dye laser

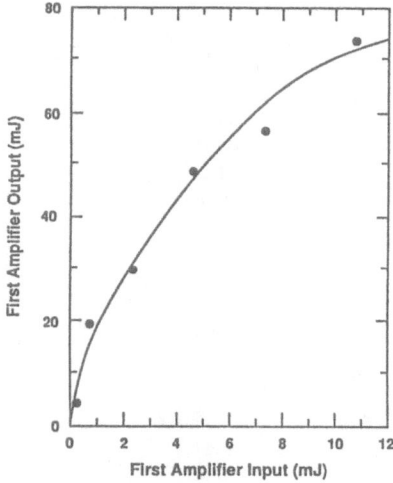

Fig. 4.25. Saturation curve for the amplifier stage for the TBS/dioxane dye laser

4.7.5 Power Oscillator Configuration

Experiments were made, using one of the amplifier stages in a power oscillator configuration, to compare its beam quality and energy output with the MOPA configuration just discussed. The optical layout of the power oscillator is given in Fig. 4.26.

The results were beyond expectations and pointed to the possibility of an even more compact laser system. The MOPA design has the advantage of potentially excellent beam quality, but exploring the performance of a power oscillator to see if sufficient beam quality and output could be achieved in a single stage was also of interest. Clearly there were major advantages in the areas of system reliability and lower maintenance and in having a single oscillator stage that required two instead of three excimer pump lasers.

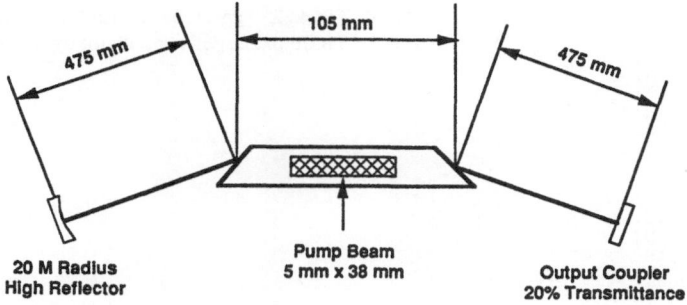

Fig. 4.26. Power oscillator optical cavity configuration

4.7.6 Power Oscillator Performance

Measurements of beam divergence and energy were made for the optical configuration in Fig. 4.26. The performance data is given in Table 4.11, and the spatial profile of the output beam for both the horizontal and vertical directions is given in Fig. 4.27.

4.7.7 Homogenized Pump Beam Experiments

Homogenizers were used to smooth the pump beams. The XMR homogenizers used produce a very flat intensity profile, both vertical and horizontal, in a plane about 7–10 cm from the exit window of the homogenizer. This design allowed much better control of the pump beam uniformity and decreased the peak fluence. The configuration is shown in Fig. 4.28.

Table 4.11. Power oscillator performance data for TBS/dioxane

Best performance data measured:	
Output energy	153 mJ
Wavelength	400 nm
Efficiency	26.9%
Excimer pump energy	569 mJ
Output coupler	90% trans.
Full reflector mirror radius	20 m
Optical cavity length	1.2 m
Excimer pump beam size	3×45 mm^2
Beam divergence Horizontal	3.9 mm mR
Beam divergence Vertical	10.1 mm mR
Dye concentration. TBS in Dioxane	34 mg/l
Dye flow rate[a]	1.9 l/s

[a] Minimum used in these tests. (Note: for a clearing ratio of 2, we compute that 250 Hz and a flow volume of $4 \times 3 \times 45$ mm^3 would only require a flow rate of 0.4 l/s.)

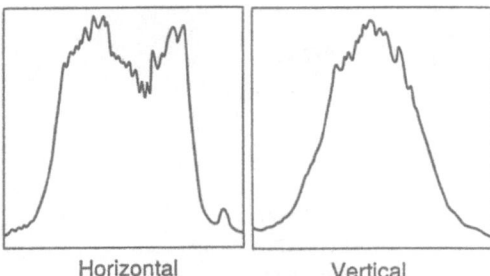

Horizontal   Vertical

Fig. 4.27. Typical beam profiles for the TBS/dioxane power oscillator dye laser

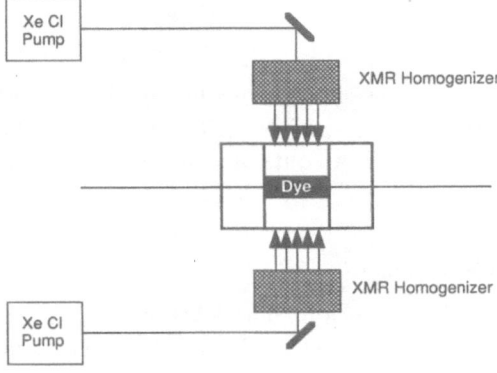

Fig. 4.28. Power oscillator excimer pumping method layout using beam homogenizers

Optical pumping was simplified by directing each excimer beam to each side of the cell. The XMR homogenizer requires angular alignment of the input beam in the vertical direction, and although not so critical in the horizontal direction, the beams must be on axis. The depth of the region of uniform intensity was about 2 cm, which was adequate since the dye cell width was only 0.5 cm. The two-sided pumping with each excimer laser beam split was done originally to average the energy on each side of the cell. We found that, if either laser dropped much in output power, the dye laser output dropped and thus required adjustment. Timing errors between the two pump lasers would also cause the dye beam to change in spatial shape and divergence. The dye-beam energy would shift to the side of the first laser that fired, and a shift in timing of only 2 ns would make a measurable change in the dye-beam profile. However, with either pumping method, the timing needed to be held to a few nanoseconds for the control of energy and beam quality.

The use of homogenizers improved the performance of the dye laser output and beam quality and further reduced the peak fluence, which reduced damage to the windows.

4.7.8 System Considerations

Several other major components are required to allow reliable operation of a dye-laser system. The dye-pumping system was certainly a critical component. The choices of pump and plumbing, as well as sufficient instrumentation and controls, are critical to successful operation. The need for alarms and fault indicators is obvious when pumping highly flammable fluids at high pressure (4 atm) and at high flow rates (3 l/s). In addition to flow controls, display of the status of the excimer laser's interlocks in the control room is essential for prudent supervision of the system. Monitoring the lasers' electrical interlocks with fiber optic links to an interlock display will allow failures to be determined quickly without leaking electromagnetic noise from the laser body. In addition to monitoring the status of the excimer lasers, it is important to monitor output of each dye cell for output energy.

References

4.1   R.S. Hargrove, Tehmau Kan: IEEE J. Quantum Electron. **QE-16**, 1108 (1980)
4.2   D. Klick: Industrial applications of dye lasers, in *Dye Laser Principles*, ed. by F.J. Duarte, L.W. Hillman (Academic, New York 1990) pp. 345–412
4.3   R.A. Tennant, M.C. Whitehead, C.R. Tallman, R.W. Basinger: High average power excimer-laser pumped dye oscillators, in Proceedings of the International Conference on Lasers '88, ed. by R.C. Sze, F.J. Duarte (STS, McLean, VA 1989) pp. 420–424
4.4   Francis Bos: Appl. Opt. **20**, 3553 (1981)
4.5   O. Uchino, T. Mizunami, M. Maeda, Y. Miyazoe: Appl. Phys. **19**, 35 (1979)
4.6   L.A. Bloomfield: Opt. Commun. **70**, 223 (1989)
4.7   V.S. Antonov, V. Brackmann: Efficiency of laser dyes under UV-excimer-laser pumping, Lambda Physik GmbH Hans-Bockler-Str. 12 D-3400 Göttingen, FRG
4.8   P. Cassard, R.S. Taylor, P.B. Corkum, A.J. Alcock: Opt. Commun. **38**, 131 (1981)
4.9   H. Gusten, M. Rinke, Chenheng Kao, Y. Zhou, M. Wang, J. Pan: Opt. Commun. **59**, 379 (1986)
4.10 W. Huffer, R. Schieder, H. Telle, R. Rave, W. Brinkwerth: Opt. Commun. **33**, 85 (1980)
4.11 A.N. Fletcher, D.A. Fine, D.E. Bliss: Appl. Phys. **12**, 39 (1977)
4.12 A.N. Fletcher: Appl. Phys. **B31**, 19 (1983)
4.13 R.H. Knipe and A.N. Fletcher: J. Photochem. **23**, 117 (1983)
4.14 V.S. Antonov, K.L. Hohla: Appl. Phys. **B30**, 109 (1983); Appl. Phys. **B32**, 9 (1983)
4.15 Radiant Dyes Chemie Gallard-Schlesinger Chemical Co., 584 Mineola Ave., Carle Place, New York
4.16 K.L. Hohla: Excimer-pumped dye lasers – the new generation, in Laser Focus, June 1982, pp. 67–74
4.17 J.T. Kunjappu, K.N. Rao: Indian J. Chem. **26A**, 453 (1978)
4.18 P. Cassard, P.B. Corkum, A.J. Alcock: Appl. Phys. **25**, 17 (1981)
4.19 Barbara F. Smith: Cyclohexane irradiation studies, Los Alamos National Laboratory memorandum CLS-1/87-330-BFS, May 27, 1987
4.20 S. Dellonte, E. Gardini, F. Barigelletti, G. Orlandi: Chem. Phys. Lett. **49**, 496 (1977)
4.21 G. Orlandi, L. Flamigni: Radiat. Phys. Chem. **21**, 113 (1983)
4.22 Barbara F. Smith: TBS-laser dye, Los Alamos National Laboratory memorandum CLS-1/87-522-BFS, December 15, 1988

4.23 R.A. Tennant, C.R. Tallman, D.E. Watkins, J.F. Figueria: High power excimer laser pumped dye laser, in Technical Digest of the International Conference on Lasers '86 (STS McLean, VA 1986) p. 18

4.24 B.R. Mauro, Stephen R. Foltyn, Virgil Sanders: Damage to fused silica windows while under simultaneous exposure to flowing solvents and laser radiation at 308 nm, Los Alamos National Laboratory document LA-UR-88-1094 presented at Symposium of Optical Materials for High Power Lasers, Boulder, Colorado, October 28, 1987

4.25 G.W. Butler: Fluid dynamic contributions to dye laser optical damage, Energy International, Inc., report E1651R058, June 1986 (unpublished)

4.26 C.R. Tallman: A study of excimer laser preionization techniques, Los Alamos National Laboratory document LA-UR-79-1630, presented at the Optical Society Excimer Laser Meeting

4.27 C.R. Tallman: Preionization techniques for discharge lasers, in Pulse Power for Lasers II, SPIE **1046**, 2–14 (1989)

4.28 R.A. Tennant: Laser Focus **17**, 65 (1981)

4.29 R.A. Tennant: XeCl laser chemical problems, in Proceedings of the International Conference on Lasers '80, ed. by C.B. Collins (STS, McLean, VA 1981) pp. 664–670

4.30 R.A. Tennant, Robert G. Anderson, Albert A. Menegat: Long life, maintenance-free excimer laser, Los Alamos National Laboratory document LA-UR-88-1003, presented at CLEO '88, Anaheim, California, 1988

4.31 R.A. Tennant: Excimer laser gas chemistry, Los Alamos National Laboratory document LA-UR-90-1018

4.32 Gregory Jursich, Denis Rofin, William Von Drosek, John Reid, Tom Frotins: Chemistry studies improve excimer gas lifetimes, in Laser Focus World, June 1989, pp. 93–96

4.33 William Hans, Peter Scott: Preventing contamination of excimer laser gases, in Laser Focus/Electro-Optics, October 1983, pp. 87–92

5. High-Power Dye Lasers Pumped by Copper Vapor Lasers

C.E. Webb

With 25 Figures

Continuous-wave dye laser systems pumped by Ar^+ or Kr^+ lasers are capable of output powers in the range of a few watts, but they operate at very low overall conversion efficiencies (typically 10^{-4} or smaller). Pulsed dye laser systems pumped by frequency-doubled Nd:YAG lasers, excimer lasers, N_2 lasers or flashlamps usually operate at higher efficiencies (typically a few times 10^{-3}), and are capable of generating tunable laser radiation with peak powers of a few megawatts. However the repetition rates of these lasers have, until very recently, been restricted to values in the 10–100 Hz range. For many applications such a low pulse repetition frequency is a serious handicap.

Copper vapor lasers (CVLs) possess unique advantages as sources of pulsed radiation for pumping dye lasers. They operate simultaneously at two wavelengths – 510.6 nm (green) and 578.2 nm (yellow) – producing pulses of 10–50 ns duration with energies of several mJ per pulse, at repetition rates in the range 2–32 kHz. The overall efficiency of commercially available CVL systems is typically 0.5–1%, so that an air-cooled CVL with a mean output power of 15 W can draw the required 3 kW from a domestic single phase electricity supply. Water-cooled versions demand only about a tenth of the cooling water supply of typical ion lasers despite the fact that commercial CVLs are typically more than twice as powerful. The most common power output range for CVLs is 40–60 W, but units giving 100 W are commercially available. Table 5.1 lists some of the properties of CVLs manufactured by Oxford Lasers Ltd.

The fact that the pump wavelengths are in the visible region of the spectrum implies that the fundamental tuning range of a CVL-pumped dye laser is restricted to wavelengths longer than 530 nm. Of course, using pump radiation of relatively low photon energy brings the corresponding advantage that the longer wavelength dyes are subject to much less rapid photolytic degradation than when they are pumped by an ultraviolet laser. This, together with the CVL's good overall efficiency, can be an important consideration in the choice of the primary pump source for any industrial scale photochemical process involving the use of tuned laser radiation.

It was not, however, just the advantages of pump laser efficiency and dye lifetime that led to the choice of the CVL as a pump laser for the atomic vapor laser isotope separation (AVLIS) process pioneered in the USA at the Lawrence Livermore National Laboratory (LLNL) in the early 1970s. Another feature of prime importance which led to this selection was the ability of the CVL to be used as an amplifier with even better extraction characteristics than it possesses

Springer Series in Optical Sciences, Vol. 65
High-Power Dye Lasers   Editor: Francisco J. Duarte
© Springer-Verlag Berlin Heidelberg 1991

Table 5.1. Copper vapor laser parameters

Model No.	ACL25	ACL45	ACL100
Average power [W]	25	45	100
Beam diameter [mm]	25	42	60
Max pulse energy [mJ]	4	9	20
Repetition frequency			
Standard range [kHz]	8–20	5–8	4–6
Extended range [kHz]	2–32	3.5–20	4–10
Pulse width[a] [ns]	15–80	20–60	30–65
Power consumption [kW]	4.5	6.5	14

[a] Depending on repetition frequency selected.

as an oscillator. Thus when the beam from a master CVL oscillator passes down a chain of amplifier units, it extracts more power than could be extracted from the individual units used as oscillators. In this way CVL beams with several kilowatts mean power can be generated.

Even at power levels of 10–50 W, CVL-pumped dye lasers find applications in which they possess worthwhile advantages over other technologies. For example, in the medical applications described in Sect. 5.6.1, the air-cooled CVL pump laser not only has the advantage of easy installation but the high gain of the accompanying pulsed dye laser design also ensures ruggedness and stability.

There are several scientific applications, such as amplification of femto-second pulses (Sect. 5.6.2) and resonance ionization studies (Sect. 5.6.3) to which the CVL-pumped dye laser is also extremely well suited. Here it is the high data rate which is the important feature, since the energy per pulse provided by these systems is usually more than adequate to ensure a good signal.

In order to understand how the capabilities of CVL pumped dye lasers arise it is first necessary to explore in some detail the mechanisms and practical considerations affecting the performance of the CVL itself.

5.1 The Elemental Copper Vapor Laser

5.1.1 History of Development

In 1965 *Fowles* and *Silfvast* [5.1] reported laser action on the 723 nm transition of neutral lead excited in a pulsed discharge containing lead vapor and a rare gas buffer. The lead vapor laser proved to be the forerunner of a new class of "self-terminating" metal vapor lasers of which the CVL and the gold vapor laser (GVL) are now the best known examples.

Laser oscillation on the green (511 nm) and yellow (578 nm) transitions of atomic copper was first demonstrated by *Walter* et al. [5.2] in 1966. The features which characterise the operation of the CVL and all other members of the class of "self-terminating" laser systems arise from the particular arrangement of energy levels in the atomic laser species (Sect. 5.1.2).

The earliest CVLs were heated by an external furnace in order to vaporize small pieces of metallic copper distributed along the floor of the laser tube. Although such devices could not be run for extended periods, these experiments established that the CVL was capable both of good efficiency and high repetition rate.

A significant step forward in CVL technology came in 1972 with the announcement by *Isaev* et al. [5.3] of the development of CVLs in which the gas discharge itself provided the heat necessary to maintain the CVL tube at its working temperature. The same group at the Lebedev Institute continued development of self-heated CVLs and reported [5.4] the attainment of power levels over 40 W in 1977 although the operating lifetime of these early devices tended to be rather short. During the same period, the US AVLIS research program at LLNL had identified the CVL as a promising pump source for the high power dye lasers used to selectively excite and ionize the atoms of the fissionable uranium isotope ^{235}U. Joint efforts by workers at LLNL and the General Electric Company led to the development of self-heated 15 W CVLs for the Venus Laser System [5.5] which came into operation in 1978. Despite these efforts the output power of practical CVLs seemed destined to remain at the 20 W level. However, the outlook changed dramatically in 1978 with the demonstration [5.6] by *Smilanski* and co-workers in Israel that, contrary to earlier expectations, high repetition rate self-heated CVLs could be operated in discharge tubes of internal diameter as wide as 40 mm. With this key demonstration of volumetric scalability the stage was set for further development of high power CVLs in the US program at LLNL. By 1980 *Anderson* et al. [5.7] had demonstrated a CVL at Livermore that produced over 110 W of average power. Amplifier units capable of generating "several hundred watts" each were reported [5.8] to be available at LLNL by 1985.

5.1.2 Excitation Mechanisms

As shown in Fig. 5.1, the lowest configuration of the copper atom has one $4s$ electron outside a complete subshell of ten $3d$ electrons, giving rise to the $^2S_{1/2}$ ground state. Just as in the case of alkali atoms, which also have a single s electron outside complete shells, there is a sequence of excited levels arising from configurations in which the outer electron is promoted into orbitals of successively higher n and l values. In particular, the first such configuration $3d^{10}4p$ gives rise to a pair of levels $^2P_{3/2}$ and $^2P_{1/2}$ which are optically connected to the $3d^{10}4s\,^2S_{1/2}$ ground state by the strong resonance transitions at 325 and 328 nm.

However, the copper atom's energy level diagram shows a feature lacking in those of alkali atoms. In Cu, the configuration $3d^9 4s^2$ has an energy well below the ionization energy, indeed it gives rise to $^2D_{5/2}$ and $^2D_{3/2}$ levels only 1.4 and 1.6 eV above the ground state. The ^2D levels are fully metastable because optical decay from them to the ground state is completely forbidden to electric dipole

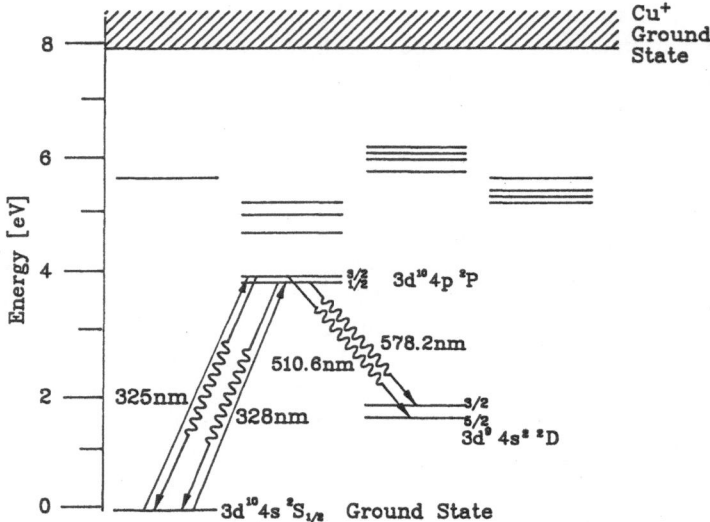

Fig. 5.1. Partial energy level diagram of atomic copper

radiation since they have the same (even) parity as the ground state. Moreover, if the central field approximation were a completely valid description of the Cu atom, the ^2P to ^2D transitions would also be forbidden since they would require *two* electrons to jump simultaneously. However, enough configuration interaction is present to provide moderate transition probabilities on the two laser transitions:

$$3d^{10} 4p\,^2P_{3/2} \;\rightarrow\; 3d^9\,4s^2\,^2D_{5/2}, \quad 511\ \text{nm (green)}, \quad (A_{21})^{-1} = 450\ \text{ns} ,$$

$$3d^{10} 4p\,^2P_{1/2} \;\rightarrow\; 3d^9\,4s^2\,^2D_{3/2}, \quad 578\ \text{nm (yellow)}, \quad (A_{21})^{-1} = 600\ \text{ns} ,$$

where A_{21} is the Einstein coefficient for spontaneous emission from the upper level 2 to the lower level 1.

Under conditions of practical interest for CVL operation, the density of copper in the vapor phase is sufficiently high (10^{15} atoms cm^{-3}) that radiation on the $^2P \rightarrow {}^2S$ resonance lines is completely trapped. The transitions from the ^2P to the ^2D levels (i.e. the laser transitions) then represent the only effective radiative decay routes for atoms in the upper laser levels.

Because the decay from the lower laser levels can occur only via collisional processes which are effective only over relatively long timescales the CVL, in common with all laser systems of the self-terminating category, completely fails to satisfy the necessary condition for cw operation

$$(A_{21})^{-1} > (g_2/g_1)\tau_1 , \tag{5.1}$$

in which g_1 and g_2 are the respective statistical weights of upper and lower levels, and τ_1 is the effective decay time of the lower laser level under conditions of laser operation.

Although the cross sections for collisional excitation of atoms by electrons do not follow rigorously the optical selection rules (especially at energies near the threshold for excitation) it is still true to say that collisional processes which involve the promotion of an *inner* electron tend to be of low cross section, while those corresponding to fully allowed optical excitation of the *outer* electrons are likely to have higher cross section values. Hence, if a fast-rising pulse of current is applied to a discharge tube containing a mixture of buffer gas (e.g. neon at 20–100 Torr) and copper vapor (0.5–1 Torr), the rate of collisional excitation of the ^2P upper laser level will exceed that of the ^2D lower laser levels provided that the electron temperature remains above about 1.2 eV. Thus, in the first 50 ns or so of the discharge current pulse (which might typically last for 300–500 ns) a substantial population inversion can be created.

The gain on the 511 and 578 nm transitions from ^2P to ^2D is sufficiently strong ($5 \times 10^4 \, \mathrm{m}^{-1}$ at 511 nm) that oscillation builds up in a very small number of cavity round trips. By the time this has occurred, stimulated emission is dumping population into the ^2D metastable lower levels which have no means of disposing of it on the timescale of 10 ns. This precipitates a rapid switch in the behaviour of the active medium – from high gain to high loss – and the laser action self-terminates well before the end of the current pulse. The fact that the total cross section for inelastic energy loss by electrons is dominated by the process leading to excitation of the upper laser levels, coupled with the good quantum ratio for the laser transition, is responsible for the high efficiency of the CVL.

5.1.3 Practical Considerations

Figure 5.2 shows a schematic diagram of the construction of a typical longitudinally excited CVL. The discharge tube (a ceramic tube typically of 20–40 mm

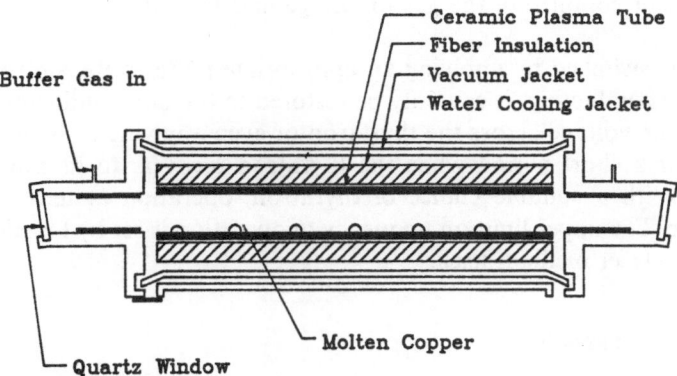

Fig. 5.2. Copper vapor laser discharge tube: schematic

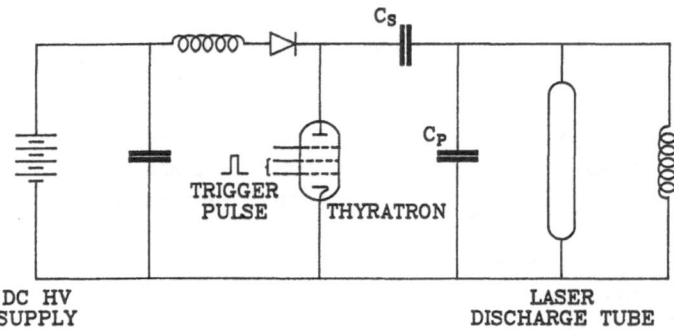

Fig. 5.3. Copper vapor laser: high voltage circuit

internal diameter and 1 m length) is surrounded by an insulating layer of ceramic fibre material for thermal insulation. The entire discharge tube and thermal insulator assembly is contained within a gas envelope of pyrex or silica closed at each end by silica windows set at Brewster's angle or tilted a few degrees from the normal to the laser axis. One of the two cylindrical electrodes at either end of the discharge tube is electrically connected to a metal tube enclosing the discharge tube which acts both as a heat sink (to absorb the heat radiated by the discharge tube) and as a coaxial current return path. Small pellets of copper are placed at intervals down the length of the discharge tube which also contains a buffer gas, neon or helium, at a pressure of 10–100 Torr.

The circuit which powers the laser is shown in Fig. 5.3. It comprises a high voltage supply capable of resonantly charging the storage capacitor C_s to a voltage between 10 and 20 kV. When the thyratron fires, the charge on the storage capacitor is transferred to the peaking capacitor C_p mounted close to the laser tube assembly and connected directly across the two electrodes. As the voltage on the peaking capacitor rises, the gas in the main discharge tube undergoes breakdown and the peaking capacitor rapidly discharges through the low inductance circuit comprising the gas discharge and the external coaxial current return path.

The thyratron is switched by applying an appropriate pulse to its control grid. However, once conducting, it can only be restored to the non-conducting state by removing the voltage across the thyratron or even applying a slightly negative voltage for a short period while the thyratron recovers to its non-conducting state. With a suitable choice of thyratron, operation at tens of kilohertz is possible. The upper limit on frequency of operation is set by kinetic processes within the laser medium itself.

5.1.4 Repetition Rate Capability

Because both ^2D lower levels of the laser transitions in copper are metastable, with no allowed optical decay route to the ground state, substantial population remains in these levels after the excitation pulse and the laser pulse have terminated. It was originally believed that the only effective mechanism for depopulating these levels was via diffusion of ^2D atoms to the tube wall and de-excitation in wall collisions. It is indeed possible that this mechanism is effective at low buffer gas pressures in tubes of very small diameter (less than 10 mm). However, in 1978 *Smilanski* [5.6] demonstrated that even at high buffer gas pressures (exceeding 100 Torr of neon) in tubes of 40 mm diameter, the CVL could still be operated efficiently at kilohertz repetition rates. This fact clearly implies that some mechanism acting throughout the volume of the discharge tube must be effective in depopulating the ^2D lower laser levels in the interval between successive laser pulses.

Studies at LLNL [5.9] and Oxford [5.10] have identified the mechanism responsible for deactivating copper atoms in the ^2D levels as superelastic collisions of plasma electrons. Experiments monitoring the decay of lower laser level population immediately after the laser pulse and throughout the interval between successive laser pulses shows that the ^2D population undergoes an initially rapid decay (time constant of order 10 µs) and subsequently a slower decay (time constant of order 100 µs). Calculations indicate that the population of ^2D levels is at all times mediated by electron collisions, both excitation collisions with ground state copper atoms populating the ^2D levels and super-elastic collisions transferring population from the ^2D levels back to the ground state. The population in the ^2D levels is therefore in quasi-equilibrium with that of the ground state at the electron temperature.

The initial rapid decay of ^2D population is associated with the decay of the electron temperature towards the ambient buffer gas temperature. The slower decay is associated with the relaxation of the buffer gas temperature itself. The thermal diffusivity of the buffer gas is sufficient to allow adequate relaxation of

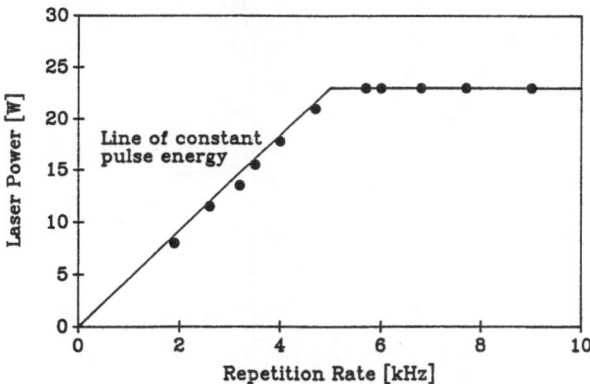

Fig. 5.4. CVL output power vs. repetition frequency at constant voltage (discharge tube diameter 42 mm.) From [5.10]

the ^2D population on the 100 µs timescale to enable the next excitation pulse to produce an inversion with respect to the remaining lower laser level population.

It is therefore the thermal relaxation time of the buffer gas which has the dominant effect in determining the optimum repetition rate of the CVL. In discharge tubes of small bore (diameter less than 20 mm), optimum repetition rates for CVL operation tend to be of order 10 kHz or even higher. As the diameter of the tube is increased the optimum repetition rate tends to fall, so that for a 40 mm tube the optimum repetition rate is in the range 6–7 kHz and for tubes of 60 mm bore it is 4.5–5.5 kHz.

As shown in Fig. 5.4, for a tube of diameter 42 mm the energy per pulse is constant at frequencies up to 5 kHz so that the output power depends linearly on frequency. Above 5 kHz however, the energy per pulse decreases with the increasing repetition rate so that the overall output power remains more or less constant over the range up to 8–9 kHz after which it begins to decrease rapidly.

5.1.5 Optimum Temperature and Green/Yellow Ratio

The optimum temperature for CVL operation represents a compromise between two conflicting tendencies. As the operating temperature is increased the vapor pressure of copper increases rapidly, making more ground state copper atoms available for excitation to the upper laser level. However, as the ground state copper density increases, the mean energy of the electron energy distribution during the excitation current pulse undergoes a corresponding decrease because the electrons need less energy to create a given rate of ionization. The result of

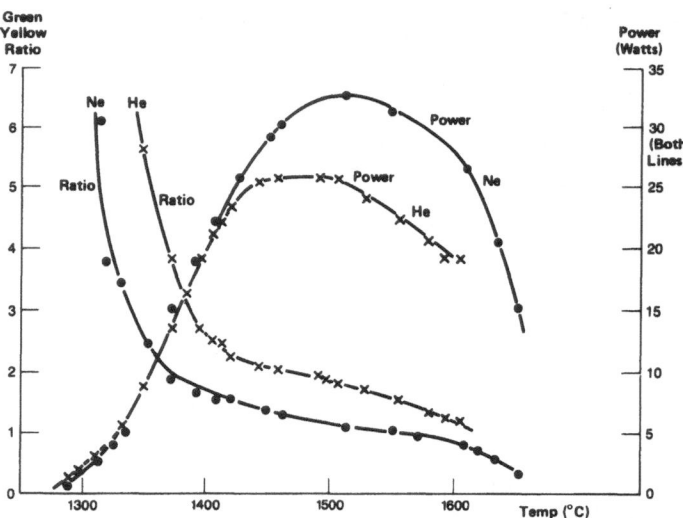

Fig. 5.5. CVL output power and green: yellow power ratio vs. tube temperature for helium and neon buffer gases. (discharge tube diameter 42 mm.) Courtesy Oxford Lasers Ltd.

this decrease in electron mean energy is that the ^2P upper laser levels are no longer collectively excited in preference to the ^2D lower laser levels, a tendency which increases rapidly as the electron mean energy falls below about 1 eV.

Because of the difference in energy of the ^2D lower laser levels, and the differing gain on the two transitions, the optimum temperature for the green laser transition is somewhat lower than that for the yellow laser transition. As shown in Fig. 5.5, as the CVL is heated up from cold, oscillation on the green transition starts when the tube temperature reaches 1260°C and reaches its peak power when the tube temperature is 1500°C. The yellow transition does not start to oscillate until the tube temperature reaches 1300°C but, with neon buffer gas, output power on the yellow transition continues to increase until a tube temperature of 1600°C is reached, after which it too shows a decline with increasing temperature. With a CVL of 40 mm diameter operating under conditions for optimum power output (repetition rate 6.5 kHz, temperature 1520°C, buffer gas neon) the ratio of output power on the 511 nm transition to that on the yellow 578 nm transition is about 1.2:1. Operation at temperatures above this optimum will shift the ratio in favour of the yellow transition.

5.1.6 Power Scaling

Once it had been established [5.6] that the output power of the CVL could be scaled in tube diameter, the problem of constructing CVL modules for use in high power amplifier chains centered on the technology of maintaining refractory discharge tubes (larger than 50 mm in diameter and 1.5 m in length) at elevated temperatures for extended periods of operation. Because of the decrease of optimum repetition rate with increasing tube diameter described in Sect. 5.1.4, the power output capability does not scale in linear proportion to the tube volume. However, increasing the tube diameter and the length of the tube both lead to increases in the output power capability of the CVL amplifier. The power capability of an individual CVL module is determined more by technological considerations than by fundamental ones, at least up to 500 W per unit.

For industrial scale installations involving many CVL amplifier units, considerations of ease of maintenance and availability at all times of a large percentage of the designed output power capability imply that a unit size somewhere in the 200–400 W range may be optimum. Individual amplifier units can be removed for maintenance purposes from the chain. While a replacement unit is substituted, the beam is diverted around the space belonging to the missing unit, so that the output power available at the end of the chain suffers only a small (e.g. 10%) decrease until a new unit is substituted.

5.2 Copper Halide Lasers

In the early 1970s the technological problems associated with running discharge tubes continuously at the high temperatures required by CVLs employing elemental copper caused attention to be focused on alternative CVL laser media which could operate at lower temperatures. A CVL whose active medium comprised a rare gas buffer seeded with the vapor of copper iodide was reported by *Liu* in 1973 [5.11]. Later the same year a similar system employing copper bromide vapor as the laser material was reported by *Chen* [5.12]. With these systems, adequate vapor pressures of the copper halide could be obtained in discharge tubes maintained at temperatures as low as 500–600°C. Since, however, copper atoms can re-associate with halogen atoms on the wall of the discharge tube in the interval between laser pulses, in order to maintain a substantial fraction of the copper halide molecules in dissociated form it is necessary to operate copper halide lasers within a limited range of pulse repetition frequencies. If the pulse repetition frequency is too low, the interval between pulses allows a substantial fraction of the atomic copper to recombine to the halide form. If the repetition rate is too high, then the population remaining from the previous pulse in the ^2D lower laser levels leads to a decrease in the pulse energy, as in the case of the elemental copper laser (Sect. 5.1.4).

Because the operating temperature range of the halide CVL lies well within the range at which silica has satisfactory mechanical properties the problems of tube design and construction are greatly eased. Glass-blowing techniques can be used to construct the silica vessels for halide CVLs, permitting much more complex geometries in discharge and electrode configurations to be manufactured than is the case for the elemental CVLs where the choice of construction materials is limited to refractory ceramics.

There are however, some disadvantages in the use of halide molecules as the laser material for CVLs. Firstly, the halogen atoms liberated in the dissociation of the halide molecules give rise to a significant population of electron attaching species in the discharge. This can lead to an undesirable tendency for the discharge to constrict in cross sectional area so that the gain medium does not fill the tube aperture uniformly. Studies carried out by *Sabotinov* and co-workers [5.13] have shown that this problem can be overcome by using discharge tubes containing internal apertures so that the discharge is not allowed to contact the walls of the tube. An additional advantage is that the spaces between the annular disc apertures can serve as reservoirs for copper bromide.

One problem of copper bromide lasers is associated with the build up of copper bromide on the optical windows at the end of the discharge tube. If the windows are cold, migration of the halide builds up an optically absorbing layer on the end windows, a problem which can be temporarily overcome by heating the windows. If, however, the windows are maintained continuously at high temperature there is a tendency for the material to be deposited in the form of metallic copper which adheres to the window very tenaciously. The problem of window contamination by copper is also present in elemental CVLs but the much higher volatility of the halide exacerbates the problem considerably.

In CVLs filled with high purity CuBr and neon buffer gas only, Sabotinov notes that the laser output is confined to the annular region of the discharge tube with very little gain at the centre of the tube. By adding hydrogen to the discharge he finds that the output becomes not only much more uniform across the tube cross section but is increased by more than a factor of two over discharge tubes not containing hydrogen. Progress in the development of copper halide and other metal vapor lasers in Bulgaria and the Soviet Union is detailed in a recent book edited by *Petrash* [5.14].

The pulse durations for the halide lasers are considerably longer than those of elemental CVLs of the same output power. This allows the optical cavity to exert stronger control over the output beam quality than is the case with elemental CVLs.

At the time of writing (1990), the highest output powers reported for halide CVLs are in the range of 20 W. Scaling to higher output power by increasing the discharge tube diameter may be more difficult in halide CVLs than in elemental CVLs. This is because, as the diameter increases, the repetition rate limitation set by lower laser level decay may shift the optimum pulse repetition frequency to values so low that an adequate population of atomic copper cannot be maintained in the face of rapid re-formation of copper halide molecules in the interval between pulses.

Nevertheless, halide CVLs do possess several attractive features not least of which is their short warm-up time.

5.3 CVL Beam Quality

5.3.1 The Importance of Beam Quality

For a CVL-pumped dye laser to operate successfully at the high pulse repetition rate (6–10 kHz) of the CVL, the active volume of the dye laser must be cleared of used dye solution between shots. A very rapid dye flow rate is therefore required. The limitation on flow velocities which can be achieved in practice imposes a limit on the height of the region of the dye excited by the CVL beam in the direction of dye flow. This, together with the need to excite the dye in the focal region as strongly as possible, imposes a requirement on the beam divergence of the CVL pump laser since the linear dimension of the smallest focal spot achievable with a laser beam of full angle divergence θ is $f\theta$, where f is the focal length of the focusing lens.

In many laser systems (e.g. Ar^+ or HeNe) the gain is sufficiently low that only stable cavity configurations can be used. In such cw or long pulse systems the laser beam makes many round trips of the cavity while building up to full intensity. In doing so, the lower order TEM modes dominate over higher order modes because of their lower loss. Such lasers are characterized by intrinsically low divergence output beams.

In contrast, in short-pulse high gain systems (such as the CVL, excimer lasers, and N_2 lasers), the gain period is so short that there is time enough for light to make only a few transits of the optical cavity while the medium exhibits gain. Moreover, such laser systems are characterized by a gain volume of very wide cross section. In order for power to be extracted efficiently from all points within the gain volume of such lasers it is necessary to use either a plane–plane cavity or an unstable resonator having a very high mode volume within the active medium.

5.3.2 Plane–Plane Cavities

Figure 5.6 shows schematically a CVL fitted with a plane–plane resonator.

The cavity would typically comprise one high reflectance mirror and an output coupler in the form of an uncoated silica blank. The cavity length must be kept to the minimum value consistent with the geometry of the laser head, in order not to waste any of the gain time available for building up the laser beam intensity from its spontaneous emission seed. The cavity mirror separation for a 45 W CVL is typically 2 m and the maximum duration of the gain pulse at the green and yellow wavelengths is about 60–80 ns. There is therefore only time sufficient for about five or six cavity round trips during the gain period of the laser medium.

An estimate of the beam divergence can be made by imagining a small volume near the tube centre to provide the seed radiation, in the form of spontaneous emission, from which the stimulated emission in the cavity will grow. The divergence angle of the output beam will be given approximately by the diameter of the laser tube divided by the distance travelled before the beam exits, i.e. about 5–6 times the resonator length. For a laser tube of 42 mm bore and resonator length of 2 m this implies an angle of divergence of 5 mrad. If focused by an aberration free f 4 lens, such a beam can be brought to a focal spot of minimum diameter about 1 mm.

Although it may be possible to design CVL oscillators of narrow bore (less than 25 mm say) which have sufficiently low divergence to pump dye lasers directly, for most applications involving high power CVL oscillators the divergence provided by plane–plane cavities of aperture greater than about 40 mm is simply too large for effective dye laser pumping.

5.3.3 The Unstable Cavity

An unstable cavity (sometimes called an unstable resonator) is an optical cavity with a high diffraction loss. One form of the unstable cavity frequently used for CVLs is formed by a pair of confocal mirrors as shown in Fig. 5.7. The small convex mirror and the large concave mirror form a telescopic combination whose cavity magnification factor is given by the ratio of their focal lengths.

The properties of this type of telescopic confocal cavity arrangement have been discussed extensively in the Soviet literature by *Zemskov* et al. [5.15] and *Isaev* et al. [5.16]. The action of the unstable cavity can be visualized by considering the cavity to be seeded by spontaneous emission from an arbitrary source point inside the gain medium. Spontaneous emission travels in both directions along the cavity axis. Of the radiation initially travelling towards the convex mirror some is reflected at its surface, and for its return transit appears to diverge from an image point somewhere between the surface of the mirror and its focal point. This image acts as a virtual source seeding the cavity with radiation travelling in the direction of the concave mirror. Most of this radiation exits after the first reflection but some is intercepted and reflected back by the convex mirror, once again appearing to diverge from an image point even nearer its focal point. This image now acts as the virtual source for the second transit of the cavity. Again, most of the light leaves the cavity after the second transit. The light which is intercepted by the convex mirror again forms a virtual source even closer to the focal point which acts to produce an even more collimated beam on the next transit. At each successive cavity transit the divergence angle of the light in the cavity is reduced by the magnification factor of the cavity.

If the light emerging from the cavity is collected by a lens and imaged in its focal plane the radiation which exited after the first cavity transit will be focused into a disc of large diameter, that from the second transit into a disc of smaller diameter and so on, ultimately approaching the diffraction limit of the system.

Because the time available for the light to make cavity round trips is limited by the finite duration of the gain pulse of the CVL only the radiation which exits on the third and fourth cavity transits is of very low divergence. Thus the low divergence output power may be limited to only about 50% of the power available from the same CVL unit using a plane–plane cavity resonator since only the second half of the gain pulse is effective in providing low divergence output.

A significant improvement in the efficiency of extraction of laser power in low divergence beams has been achieved in by *Naylor* et al. [5.17] with the introduction of the off-axis unstable cavity. As illustrated in Fig. 5.8 this arrangement takes advantage of the fact that, at the start of each laser pulse, gain appears near the plasma tube wall a few nanoseconds before it appears at the tube axis. Thus, light can be fed back into the cavity earlier (allowing more cavity round trips to be made within the gain period) if the small convex mirror feeds back light from the edge rather than from the axis of the plasma. In the focal plane of a lens the output from such a cavity will image to a series of rings of successively smaller diameter corresponding to the successive round trips of the laser pulse within the cavity (see Fig. 5.8). By introducing an appropriately sized aperture in the focal plane of the imaging lens, the output of the laser can be restricted entirely to low divergence light.

In such a cavity the total power coupled from the laser is approximately 75% of the power available using a plane–plane resonator. Of the total output power, 80% is available within 600 µrad divergence angle and 70% within a divergence angle of 100 µrad. Using the same laser dimensions as in Sect. 5.3.3, a 45 W laser of 42 mm tube diameter, the divergence is reduced by a factor of more than 20, and the spot size achievable with an $f4$ lens of 42 mm diameter is 17 µm.

156   *C.E. Webb*

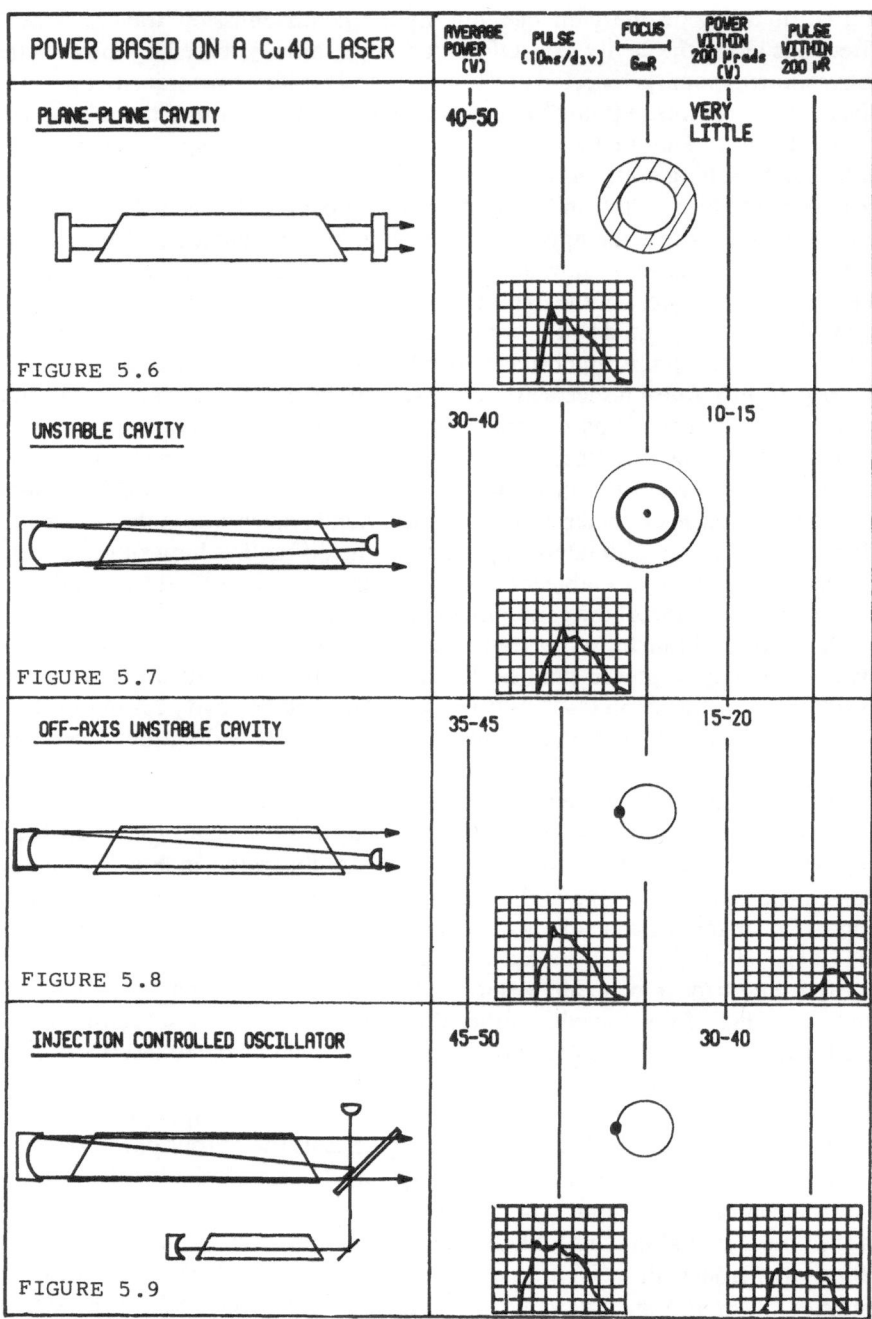

Figs. 5.6–5.9. For captions see opposite page

5.3.4 Injection Controlled Oscillator

For installations involving high power oscillator-amplifier CVL chains the radiation provided by the master oscillator must have sufficiently low divergence that it can traverse the entire length of the amplifier chain without appreciable growth in diameter. The best way of ensuring that the master oscillator produces a very low divergence beam is to use an injection controlled oscillator configuration.

As shown in the schematic diagram of Fig. 5.9, the beam from a small CVL (whose beam divergence is limited by intra-cavity apertures) is launched into a 40 W CVL module controlled by an off-axis unstable resonator. By adjusting the relative timing so that the discharge in the small laser module fires before that in the 40 W module, it is easy to ensure that the main oscillator cavity is filled with radiation of low angular divergence before its gain pulse starts to rise. The coupling of the cavities of the two laser modules is accomplished by positioning a beam splitter to enable them to share the small convex mirror as an element common to both. With this arrangement the 40 W laser can now emit low divergence light over practically the entire gain period so that it produces a mean power of 38 W of which 32 W are within a beam of 400 μrad divergence [5.18].

5.3.5 CVL Amplifier Properties

One of the most attractive features of the CVL as a pump for high power dye laser systems is its excellent performance as an amplifier. In fact, it is usually possible to extract 20% more power from a given CVL module when operated as an amplifier, than when the same module is operated as an oscillator within a plane–plane resonator. For example, a CVL module capable of providing 58 W output at 5 kHz when operated as a stand-alone oscillator is capable [5.18] of adding 70 W when used as an amplifier.

Although it is necessary to provide precise timing of the oscillator pulse with respect to the amplifier pulse in order to achieve optimum power extraction, the peak of the timing curve is sufficiently broad (± 10 ns) that the command jitter of typical CVLs (± 1–2 ns) presents no problem in maintaining timing to the accuracy required. It is also important to note that the gain pulse duration is a function of CVL tube diameter. CVL discharge tubes for modules capable of

---

Fig. 5.6. Plane–plane cavity – allows efficient extraction of power over entire gain period but laser beam has high divergence

Fig. 5.7. Confocal unstable cavity – shows lower divergence output but poor extraction efficiency. Peak power occurs only late in the gain period

Fig. 5.8. Off-axis unstable cavity – improves power extraction efficiency, but low divergence beam still occurs late in the gain period

Fig. 5.9. Injection controlled oscillator – permits low divergence beam with high output power to be efficiently extracted for entire duration of gain period. Courtesy of Oxford Lasers Ltd.

100 W are typically 60 mm in bore diameter, and gain is present for some 100–120 ns on each pulse. In order to extract power from an amplifier efficiently it is necessary that the master oscillator pulse should have a similar duration. If the master oscillator CVL operates in a laser tube of small diameter it may be necessary to provide geometric beam expansion as well as some passive optical means of stretching the pulse from the oscillator before the beam enters the amplifier in order to ensure that these conditions are fulfilled [5.19].

5.4 CVL Pumped Dye Lasers – Theory

5.4.1 Rate Equations, Gain, Saturation Intensity

The previous sections have highlighted the properties of the CVL which are important in understanding its role as a pump source for high repetition rate dye lasers. Armed with this insight we can now proceed to consider the response of the dye laser medium itself.

The most complete theoretical account of the rate equations for CVL pumped dye lasers is that published in 1980 by *Hargrove* and *Kan* [5.20]. Their treatment follows that of *Teschke* et al. [5.21] originally developed for cw and long pulse dye lasers, in terms of the kinetic processes indicated in the schematic energy level diagram Fig. 5.10.

The two types of dye laser amplifier geometry considered in this treatment, longitudinally pumped amplifiers (LPA) and transversely pumped amplifiers (TPA), are indicated in Figs. 5.11a and 5.11b, respectively.

Because the duration of the typical CVL pumping pulse (typically 10–40 ns) is long compared to τ_{21} the relaxation time (typically 10^{-12} s) of the second excited singlet level S_2 but short compared to $(k_{st})^{-1}$ the time characteristic of intersystem crossing to T_1 (the lowest triplet level), both the S_2 and T_1 populations can be neglected for any time scales of interest, and the system can be considered in quasi-steady state.

Within these approximations time derivatives of both dye laser beam and pump beam intensities as well as populations can be set to zero, yielding in the

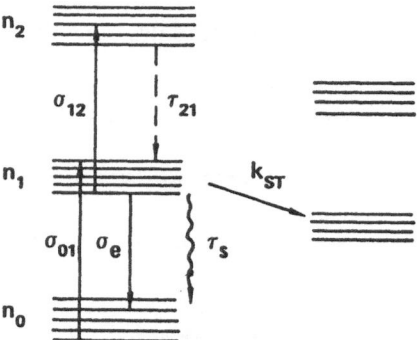

Fig. 5.10. Energy level diagram of dye molecule used in rate equation model. From [5.20]

a)

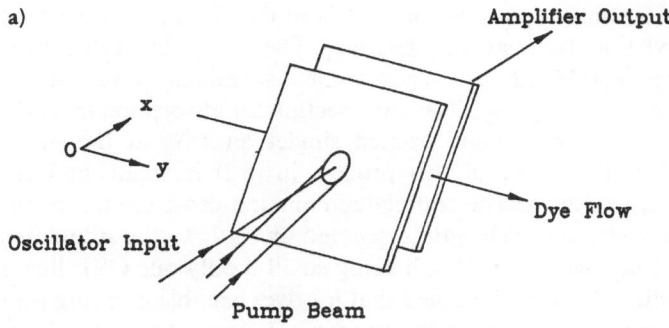

b)

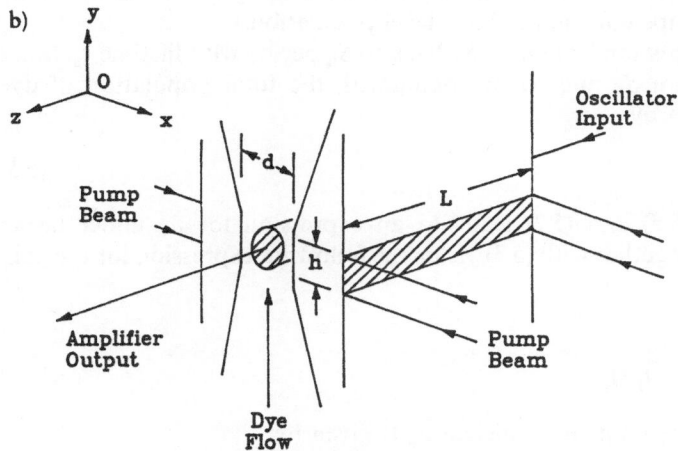

Fig. 5.11 (a) Geometry of longitudinally pumped dye amplifier (LPA); **(b)** Geometry of transversely pumped dye amplifier (TPA). From [5.20]

case of LPA geometry

$$\frac{dn_1}{dt} = 0 = \sigma_{01}(\lambda_P)n_0 I_P + \sigma_{01}(\lambda_L)n_0 I_L - (n_1/\tau_s) - \sigma_e(\lambda_L)n_1 I_L \tag{5.2}$$

$$\frac{dI_P}{dx} = -[\sigma_{01}(\lambda_P)n_0 + \sigma_{12}(\lambda_P)n_1]I_P \tag{5.3}$$

$$\frac{dI_L}{dx} = [\sigma_e(\lambda_L)n_1 - \sigma_{01}(\lambda_L)n_0]I_L \ . \tag{5.4}$$

The equations for the TPA case are similar, except that dI_L/dx must be replaced by dI_L/dz in (5.4).

In the above equations n_0 and n_1 are the populations of the ground state S_0 and first excited singlet S_1 state, respectively, while I_P and I_L represent respectively the photon fluxes (photons $cm^{-2} s^{-1}$) of the CVL pump beam at wavelength λ_P and dye laser beam at wavelength λ_L. The S_0–S_1 absorption cross section at the pump wavelength is $\sigma_{01}(\lambda_P)$ while its residual value at the wavelength of the dye laser is $\sigma_{01}(\lambda_L)$. The cross section for absorption from the excited singlet level S_1 to the second excited singlet level S_2 at the pump wavelength is $\sigma_{12}(\lambda_P)$. The effect of this process in (5.2) is small, and the corresponding term is neglected. The stimulated emission cross section of the S_1–S_0 transition at the lasing wavelength is denoted by $\sigma_e(\lambda_L)$ – the stimulated emission on S_1–S_0 at the *pump* wavelength being small if only one CVL line is used. (In this connection it should be noted that for dyes capable of lasing near 578 nm it may be necessary to completely remove any component at 578 nm from the pump beam to ensure that emission stimulated by 578 nm in the pump beam does not compete for upper laser level population.)

Spontaneous emission from level S_1 back to S_0 occurs with lifetime τ_s. Since no states other than S_0 and S_1 are populated, the total population of dye molecules n, is gven by

$$n = n_0 + n_1 . \tag{5.5}$$

Solving between (5.5) and (5.2) leads to an expression for n_1 under lasing conditions, which together with (5.3) yields the following expression for the gain coefficient α

$$\frac{1}{I_L} \frac{dI_L}{dx} = \alpha = \frac{\alpha_0}{1 + I_L/I_s} . \tag{5.6}$$

In (5.6), the small signal gain coefficient α_0 is given by

$$\alpha_0 = \frac{n\sigma_e[\sigma_{01}(\lambda_P)\tau_s I_P - \sigma_{01}(\lambda_L)/\sigma_e(\lambda_L)]}{1 + \sigma_{01}(\lambda_P)\tau_s I_P} , \tag{5.7}$$

while the saturation intensity (photon flux) is given by

$$I_s = \frac{[1 + \sigma_{01}(\lambda_P)\tau_s I_P]}{\sigma_e \tau_s [1 + \sigma_{01}(\lambda_L)/\sigma_e(\lambda_L)]} . \tag{5.8}$$

To the extent that CVL-pumped dye lasers are characterized by strong pumping (i.e. $I_P\tau_s\sigma_{01}(\lambda_P) \gg 1$) then (5.7) indicates that the small signal gain may approach the complete inversion value α_{0m}, i.e.

$$\alpha \to \alpha_{0m} = n\sigma_e(\lambda_L) . \tag{5.9}$$

In the same limit, the saturation intensity I_s becomes directly proportional to the pump intensity,

$$I_s \to \frac{\sigma_{01}(\lambda_P) I_P}{\sigma_e[1 + \sigma_{01}(\lambda_L)/\sigma_e(\lambda_L)]} . \tag{5.10}$$

Integration of the expression (5.6) in the direction of the dye laser axis (i.e. from
$z = 0$ to $z = L$ in the TPA case) leads to the following expression for the single
pass gain factor G of the dye laser of length L

$$\ln G = \ln\left[\frac{I_L(z = L)}{I_L(z = 0)}\right] = \alpha_0 L - \frac{(G-1)I_L(z=0)}{I_s} . \tag{5.11}$$

5.4.2 Longitudinal Pumping Geometry

The arrangement shown in Fig. 5.11a in which the CVL pump beam is collinear
with the axis of the dye laser amplifier cell offers the advantage that, because of
its axial symmetry, good dye laser beam uniformity may be somewhat easier to
ensure with this geometry than with a pump beam transverse to the dye laser
axis. For this reason, longitudinal pumping geometry is attractive for the
oscillator section of an all CVL-pumped oscillator-amplifier dye laser design
(Sect. 5.4.4) but this geometry has not been favoured for use in the power
amplifier sections of very high power dye laser amplifier chains.

The disadvantage of the longitudinal geometry arises, according to the
analysis of *Hargrove* and *Kan* [5.20], as a result of the effects of excited state
absorption of the pump beam. Figure 5.12 shows the results of their calculations
for a longitudinally pumped amplifier (LPA) lasing at 572 nm on a dye solution
of Rhodamine 6G (Rh6G) pumped only by the 510.6 nm CVL line.

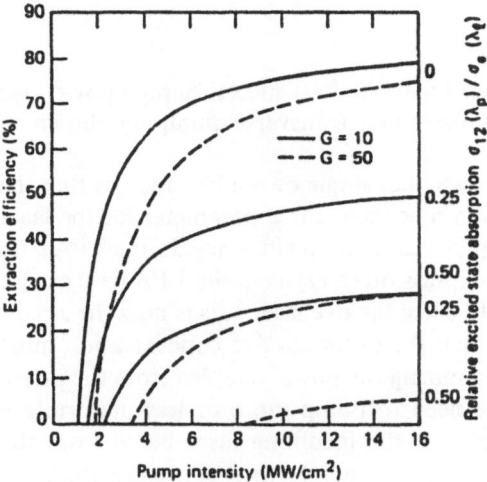

Fig. 5.12. LPA: calculated photon extraction efficiency vs. pump intensity for two values of
amplifier gain G, and two values of relative excited state absorption cross section $\sigma_{12}(\lambda_P)/\sigma_e(\lambda_L)$.
Calculations pertain to Rh6G: $\lambda_P = 511$ nm, $\lambda_L = 572$ nm, $\sigma_e(\lambda_L) = 1.86 \times 10^{-16}$ cm^2, $\sigma_{01}(\lambda_P)$
$= 1.66 \times 10^{-16}$ cm^2, $\tau_s = 4.8$ ns. Pump transmission T fixed at $\exp(-3)$. From [5.20]

The extraction efficiency η, representing the ratio of the power added to the beam at λ_L to the pump power incident on the cell at λ_P is given by

$$\eta = \frac{(\lambda_P) I_L(x = 0)}{(\lambda_L) I_P(x = 0)} (G - 1) . \tag{5.12}$$

It is evident from Fig. 5.12 that the extraction efficiency which can be achieved in LPA geometry is strongly influenced by the presence of even very small values of excited state absorption at the pump wavelength. For example an LPA of power gain $G = 10$ operates at about half its potential efficiency if the ratio of $\sigma_{12}(\lambda_P)$ to $\sigma_e(\lambda_L)$ is only 0.25.

Excited state absorption also determines the pump beam penetration depth. For the case of $\sigma_{12}(\lambda_P)$ equal to $\sigma_{01}(\lambda_P)$, it follows trivially from (5.3) and (5.5) that, independent of I_L and I_P, the depth to which the pump penetrates into the dye medium is fixed at $d_P = [n\sigma_{01}(\lambda_P)]^{-1}$. Because the pump beam cannot bleach its way into the medium, the active gain length of the LPA is restricted to values of order d_P, and the maximum value of G in this case would be given by

$$\ln G < \alpha_{0m} d_P < \frac{\sigma_e(\lambda_L)}{\sigma_{01}(\lambda_P)} . \tag{5.13}$$

If excited state absorption is small, the bleaching limitation is less important, and the LPA geometry can be successfully applied, especially in situations of low saturated gain and high ambient flux such as are typical of a dye laser oscillator.

5.4.3 Transverse Pumping Geometry

The geometry which has been employed with greatest success in high power dye laser amplifier chains is that of double-sided transverse pumping shown in Fig. 5.11b.

The advantage of TPA geometry, whether single or double sided, is that the length-to-height ratio (L/h) of the pump beam is a free parameter for the laser designer to choose in optimizing amplifier extraction efficiency. In the TPA case the efficiency η is given by the same expression (5.12) as in the LPA case except that $I_L(x = 0)$ is replaced by $I_L(z = 0)$ since the dye laser axis is now the z-axis.

If single-sided pumping is used, then the optimum dye concentration must represent a compromise between maximizing the power coupled from the pump beam into the dye medium, and the need to pump more or less uniformly a region of width equal to the diameter of the incoming laser beam from the oscillator. The analysis of *Hargrove* and *Kan* [5.20] shows that because the cross section of the pump beam can be chosen so that the pump flux at the dye is sufficiently low to ensure that

$$[\sigma_{01}(\lambda_P) - \sigma_{12}(\lambda_P)] \tau_s I_P \ll [1 + \sigma_e(\lambda_L) \tau_s I_L] , \tag{5.14}$$

while the laser beam cross section is small enough to ensure that

$$I_L \tau_s \sigma_e(\lambda_L) \gg 1 \; , \tag{5.15}$$

then the pump beam absorption varies approximately exponentially with distance from the input face according to

$$I_P(x) \simeq I_P(0) \exp[-n\sigma_{01}(\lambda_P)x] \; . \tag{5.16}$$

Increasing the dye concentration beyond the optimum value therefore has the effect of restricting the excited region to a very narrow region near the input face which is pumped very nonuniformly. Reducing it below the optimum value wastes available pump power by failing to couple it into the region of best beam overlap.

For the high power multistage amplifiers used in AVLIS applications, the beam may have to propagate over tens of meters or more, and its optical quality is therefore crucial. Two-sided CVL beam pumping is therefore necessary. The pump laser beam is split into two beams of equal flux $I_0/2$ which propagate towards one another from the input faces at $x = 0$ and $x = d$ as indicated in Fig. 5.11b. If the one-way transmission factor for the pump beam through the amplifier cell is $T = \exp(-qd)$, the pump beam flux at the axis of the TPA is

$$I_P \frac{(x = d)}{2} = I_0 \exp \frac{(-qd)}{2} \; , \tag{5.17}$$

while that at the pump beam input faces is

$$I_P(x = 0) = \tfrac{1}{2}I_0 + \tfrac{1}{2}I_0 \exp(-qd) \; . \tag{5.18}$$

The ratio of pump intensities at edge and center of the amplifier cell is therefore

$$\frac{I_P(x = 0)}{I_P(x = d/2)} = \tfrac{1}{2}[\exp(-qd/2) + \exp(qd/2)] = \tfrac{1}{2}[T^{1/2} + T^{-1/2}] \; . \tag{5.19}$$

This ratio cannot be allowed to deviate from unity by more than about 20% if beam uniformity is to be maintained. This consideration immediately sets a lower bound on T_{min} of about 0.3, implying that at most 70% of the CVL pump beam can be absorbed in the dye medium. Since the optimum height of the pumped region in the dye is equal to the diameter of the beam from the oscillator it is clear that the geometric factors and the absorption restriction set an upper bound on the possible efficiency of a TPA dye laser of

$$\eta_{max} = \frac{\pi}{4}(1 - T_{min}) \frac{\lambda_P}{\lambda_L} \; , \tag{5.20}$$

even if every pump photon absorbed contributes to the amplified beam.

In practice this implies values of η_{max} of about 55% for amplifiers of good beam quality.

5.4.4 Dye Cell Design

The thermal energy deposited in the dye laser medium by the pump beam causes changes in density and hence perturbations in refractive index of the bulk solution. These would have a very deleterious effect on laser beam quality if the dye solution were static. Because the density changes which are the main source of refractive index perturbation occur on timescales of order 1 μs, they do not disturb the medium homogeneity on the timescale of a single CVL pulse (10–40 ns) but at the normal repetition rate (4–6 kHz) of CVL pumping their effect is cumulative and severe.

A detailed study of dye medium perturbations has been carried out by the group at the Israel Nuclear Research Centre [5.22]. This showed that in a 0.5×10^{-3} M solution of Rh6G in methanol with 10% ethylene glycol, flowing through a dye cell of width 0.5 mm at velocities as fast as $18.3 \, \mathrm{m\,s^{-1}}$, strong perturbations of refractive index could be detected at *all* points downstream of the 0.2 mm × 10 mm pumped stripe (and even 1–2 mm in the upstream direction) when a 20 W CVL was used to provide one-sided excitation of the cell.

A HeNe laser beam probing the dye cell parallel to the long axis of the pumped stripe showed both angular broadening and steady deflection θ_x, from which the temperature gradient in the x-direction of Fig. 5.11b could be inferred, since

$$\theta_x = L\frac{dn}{dx} = L\frac{dn}{dT}\frac{dT}{dx} \, . \tag{5.21}$$

In order to avoid serious beam deflection problems in high power TPAs it is therefore necessary to use two-sided pumping. It is also necessary to maintain a flow velocity sufficient to remove the dye solution from the pumped volume and replace it with fresh dye solution typically three to ten times during the interval between successive CVL pulses. For a pumped volume of $0.3 \times 0.3 \times 10 \, \mathrm{mm^3}$, this implies flow velocities of order $20 \, \mathrm{m\,s^{-1}}$, and volume flow rates of order $50 \, \mathrm{ml\,s^{-1}}$. As pointed out by the Israeli workers [5.22] the flow under such conditions is characterized by Reynolds numbers well in excess of the 2000, corresponding to the onset of turbulent flow. Even for flow velocities with Reynolds numbers less than 2000 they found evidence for great enhancement of thermal diffusivity over the values expected for purely laminar flow, which they ascribe to the mixing effects of vortices generated as the dye flow enters the restricted throat of the dye cell.

For such high liquid flow velocities the dye circulation system and cell must be sufficiently robust to contain the high ambient pressures that the circulation pump generates. The flow path must also be free of sharp corners or abrupt changes of cross section which can induce cavitation in the flowing dye solution. By correctly shaping the flow duct and the dye cell with taper regions upstream and downstream of the pumped region it is not difficult to avoid such effects.

Another potential source of inhomogeneity in the dye stream can arise from air bubbles extrained in the flow, the effect of which can be exacerbated by filtering the dye solution upstream of the dye cell. The filter may cause the larger bubbles to be broken up into micro-bubbles which pass through the mesh and induce scattering and hence noise on the laser beam as they pass through the laser cell. Careful attention to filter design, and its placement within the dye circulation loop is necessary to avoid this problem.

Nearly all CVL pumped dye lasers reported in the literature to date (certainly all of those employing TPA geometry and designed for high power operation) use cells in which the dye stream is totally enclosed in a cuvette fitted with optical windows to admit the pump CVL beam orthogonal to the laser axis. Free dye jets have been successfully used in LPA geometry for applications involving femtosecond pulse amplification (Sect. 5.6.2) and in the relatively low power but efficient broad band oscillator design reported by *Sun* et al. [5.23]. One reason why the open jet stream is rather unsuitable for large scale high power installations (e.g. for AVLIS) is the fire hazard associated with focusing high power CVL beams on the highly flammable dye solution stream in air. The issue of fire hazard in large scale of CVL pumped dye laser installations has been addressed by *Hammond* [5.24].

5.4.5 Pump Beam Polarization Issues

Unless polarizing elements (such as the beamsplitter in an injection-controlled CVL oscillator) are included within the cavity, the output beam from simple CVL oscillators is unpolarized. However, particularly for chains comprising several CVL amplifier modules operating under gain saturated conditions, there is good reason to use Brewster windows and to extract power with an appropriately plane polarized beam. With an unpolarized input beam, the power loss at each amplifier stage due to reflections from the uncoated surfaces of near normal incidence windows (or the unwanted polarization component reflecting from Brewster angle windows) could amount to an economically serious wastage. For this reason, current unstable resonators designs for CVL oscillators facilitate the inclusion of the extra optical components required to generate highly plane polarized beams.

Thus, particularly for high power installations, the pump beam provided by the CVL system to the dye laser amplifiers may show a high degree of plane polarization. It is therefore relevant to consider the effect of pump beam polarization on dye laser performance.

Cw dye lasers whose cavities do not contain polarization sensitive elements usually produce a beam with the same plane of polarization as their pump laser. Such dye lasers can only be made to oscillate in other, less favourable, planes of polarization by the inclusion of a rotatable Brewster plate within its cavity.

The mechanisms responsible for this polarization preference were the subject of a detailed investigation by *Phillion* et al. [5.25]. These workers concluded that the interaction of most asymmetric organic dye molecules with the polarized pump beam radiation can be well represented by a linear electric dipole oscillator model, in which the dipole direction is along one of the molecular axes. Amongst the population of ground state dye molecules those with a large component of their dipole moment parallel to the electric field of the light wave will be able to absorb energy from the pump beam more readily than others. This selectivity in absorption leads to a predominance in the population of molecules in the upper laser level of those with a dipole moment in the direction of the pump beam electric field, and thus to the observed preference of the dye laser to produce a beam with the same polarization as the pump beam.

It might be thought that the ability of the excited molecules to rotate would rapidly destroy any memory of the direction of their dipole axis at the instant of their initial excitation. However, in the liquid phase the large dye molecules are not at all as free to rotate as they are in the gas phase. They are bombarded continuously by the smaller molecules of the solvent, so that their orientation undergoes Brownian motion, gradually diffusing away in angle from their original direction with a time constant T_{rot}. If the molecule is stimulated out of the upper laser level in a time short compared to T_{rot}, the distribution of dipole directions still retains the anisotropy associated with the original excitation. In any case, the spontaneous emission lifetime sets a limit to the time available for rotational diffusion to occur.

The experiments of *Wokaun* et al. [5.26] show that for solutions of Rh6G on low viscosity solvents, the values of T_{rot} are in the range 100–200 ps, consistent with theoretical expectations [5.27].

Although the importance of pump beam polarization may be somewhat less than in the case of long pulse or cw dye lasers more strongly influenced by triplet state absorption, it is nevertheless a significant factor in the optimization of CVL pumped dye-laser amplifiers. For transversely pumped dye laser amplifiers intended for an AVLIS chain, *Hargrove* et al. [5.28] note that in order to maximize the gain for dyes similar to Rh6G the polarization of the amplifier input and pump beams should both be perpendicular to their plane of intersection.

5.5 CVL Pumped Dye Lasers – Performance

5.5.1 Broadband Dye Laser Oscillators

As long ago as 1979, in one of the earliest studies of the pumping of dye lasers by CVL radiation, *Morey* [5.29] compared directly the performance of a dye laser oscillator pumped longitudinally by a 5 W CVL with that of a similar system transversely pumped by the same CVL. The cavity configurations he employed for transverse and longitudinal pumping are shown in Figs. 5.13 and 5.14, respectively. The dispersive elements in both cases comprised two Brewster angle prisms. Some problems with thermal lensing were noted initially, but these were eliminated by replacing the glass dye cell windows and tuning prisms with components fabricated from non-absorbing fused silica.

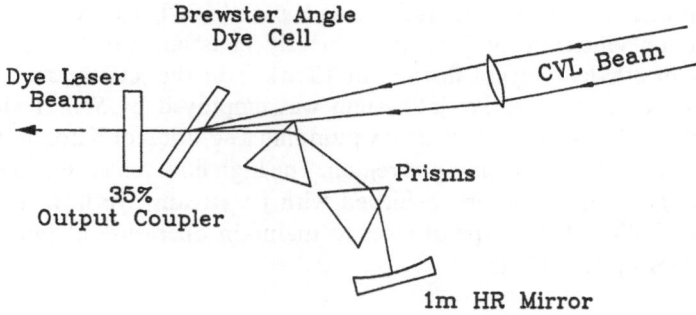

Fig. 5.13. Longitudinally pumped broadband dye laser oscillator. From [5.29]

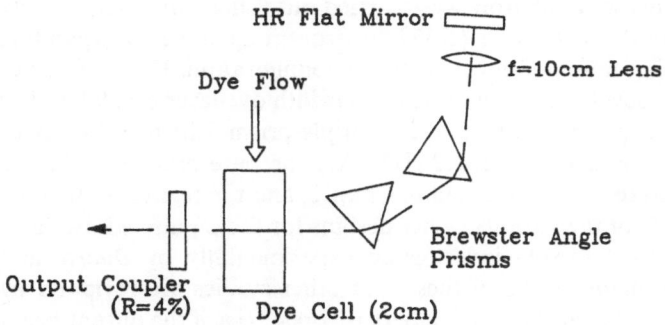

Fig. 5.14. Transversely pumped broadband dye laser oscillator. From [5.29]

Output powers of order 3 W and slope efficiencies of 59% for transverse pumping, 57% for longitudinal pumping, were achieved with Rh6G in ethanol. In addition to these impressive performance figures for conversion efficiency, Morey also noted the extremely long dye lifetimes which the absence of UV and blue/violet in the pump source spectrum made possible. The half-life of 1 liter of Rh640/ethanol solution at 10 W pump power was in excess of 560 h; in fact no degradation was noted for accumulated pump energies of $2.5 \times 10^6 \, \mathrm{J\,l^{-1}}$.

The output beam divergence characteristics that Morey observed for the two types of pumping geometry were quite different. With transverse pumping the beam was nonuniform, with large fringing fields. It had a divergence of 21 mrad in the vertical plane and 12.5 mrad in the horizontal plane. With longitudinal pumping the output beam was symmetrical, with a divergence of 1.53 mrad – very close to the diffraction limited value of 1.29 mrad. The beam profile of the longitudinally pumped version was much closer to the ideal Gaussian shape, and its far field intensity was six times greater than that of the transversely pumped version.

The good beam quality characteristic of longitudinally pumped dye laser oscillators using dye jets instead of flow cuvettes to constrain the dye stream in the pumped region was also stressed by *Huang* and *Namba* [5.30], who reported an overall conversion efficiency of 31% in a "nearly gaussian mode" for a solution of Rh6G in ethylene glycol flowing at $13 \, \mathrm{m s^{-1}}$ in the jet stream. A similar non-dispersive cavity and dye jet system was employed by *Sun* et al. [5.23] who achieved 40% conversion efficiency pumping a solution of Kiton red in glycol, flowing at $18 \, \mathrm{m s^{-1}}$ in the pumped region. The high flow velocities, and correspondingly large clearing factors, achieved with jet stream dye flow are responsible for the ability of this type of laser to maintain operation at pulse repetition frequencies up to 20 kHz.

5.5.2 Narrow-Linewidth Dye Laser Oscillators

The pioneering studies of *Pease* and *Pearson* [5.31] in 1977 on CVL pumping of narrow-linewidth dye laser oscillators were carried out using a dispersive cavity of the type introduced by *Hänsch* [5.32] incorporating a beam-expanding telescope and a grating mounted in the Littrow configuration. However, since 1979 nearly all of the development of narrow-linewidth dye laser oscillators has centred on cavity designs employing the multiple-prism Littrow (MPL) or hybrid multiple-prism grazing-incidence (HMPGI) or pure grazing incidence grating configurations discussed in detail in Chap. 2, and the references therein.

The relative merits of the various cavity designs for CVL pumped dye laser operation have been extensively investigated experimentally by *Duarte* and *Piper* [5.33, 34]. The main results of these and other studies are reviewed by *Duarte* [5.35], who points out that if the cavity is "open" (i.e. if the output beam is derived from an otherwise unused near-glancing reflection from the face of a beam-expanding prism) the efficiency can be high, but so also is the unwanted untuned ASE component of the output beam. Conversely, if the cavity is "closed" and the output is coupled via a partial reflector forming one mirror of the cavity, efficiency is lower, but the beam is considerably more free of ASE background. The pure grazing incidence scheme offers excellent linewidth properties, but tends to be somewhat more lossy. This presents more of a problem for the pump energies of 1–10 mJ per pulse characteristic the CVL than the much higher single pulse energies characteristic of low repetition rate operation with frequency-doubled Nd:YAG laser pumping.

An example of a CVL-pumped dye laser oscillator using the multiple prism beam-expander Littrow tuning technique is the design reported by *Bernhardt* and *Rasmussen* [5.36]. The cavity was defined by a flat output coupler (30% reflecting) and a diffraction grating blazed for 5th order Littrow operation in the visible. It also included a prism beam expander having a single-pass loss of 20% and a solid etalon of 5 mm spacing and 90% peak transmission. Single mode operation was achieved with 230 mW output power at 572 nm on Rh6G when the device was pumped by a 4 W of 511 nm CVL power at 6 kHz repetition rate. With electronic feedback control of the output coupling mirror position, the 60 MHz-wide beam could be scanned over a 16 GHz range without mode-hopping. However, despite its exceptionally good beam quality (near diffraction limited) for a transversely pumped system, the rather low overall efficiency (5% at the peak of Rh6G) implies that such a device would be better suited to act as the master oscillator of a MOPA system rather than as a stand-alone oscillator.

An efficiency of 20% for the dye Oxazine 17 lasing at 675 nm pumped by both green and yellow CVL outputs was observed by *Zherikin* et al. [5.37], in a transversely pumped device employing an MPL dispersive cavity. The fact that the efficiency on this dye was higher than that recorded for Rh6G (12.5%) is rather surprising, but it is perhaps significant that the material of the six-prism beam expander was a glass of very high refractive index ($n = 1.75$). The linewidth without an intra-cavity etalon was 24 GHz, but could be reduced to 1.2 GHz by the insertion of an etalon of 30 GHz free spectral range although the efficiency was almost halved. In this case the oscillator also performed best as part of an oscillator-amplifier system, which was capable of an overall 30% efficiency and 600 mW output when the pump power was divided equally between oscillator and amplifier sections.

A transversely pumped dye laser tuned by a pure grazing incidence grating scheme is described by *Broyer* et al. [5.38]. With 15 different dye solutions the tuning range covered was from 530 to 890 nm with the best efficiency of 20% being obtained with Rh6G, although peak efficiencies of 15% were typical of many of the dyes tested. In order to obtain this efficiency, however, output was coupled from the zero-order reflection from the grating (i.e. an "open" configuration). The bandwidth was 3 GHz. As usual with this coupling geometry, a rather high level of untuned ASE was present in the output beam (2–4% near the peaks of the dye tuning curves, rising to 15% near their limits). Since the main application of this stand-alone oscillator was for generating second harmonic tunable UV radiation for spectroscopic applications, the presence of this untuned background was not as bothersome as it might otherwise have seemed.

Commercially available dye laser systems, originally designed for excimer or doubled Nd:YAG laser pumping can be successfully adapted for CVL pumping. An example is the Lambda Physik FL2001 (or the FL3001 which has similar optics) the cavity configuration of which is shown in Fig. 5.15. In a significant improvement to the usual transversely pumped "open" MPL cavity configuration this laser employs the patented "LAMBDAPURE filter" which acts to suppress background ASE by deriving the output beam from a front surface reflection from the single beam-expanding prism but, instead of allowing this beam to exit directly from the cavity, it is directed back onto the grating once more [5.39]. The dispersed beam, returned from the grating at near Littrow angle, is passed through the same dye cell again to undergo amplification in a second pumped stripe before emerging as the output beam. The tuning ranges and efficiencies measured [5.40] for a Lambda Physik FL2001 dye laser pumped by a CVL (Oxford Lasers Cu40) operating at 6.5 kHz are shown in Fig. 5.16. The bandwidth of the FL2001 in these experiments was 4–7 GHz, but this could be reduced to less than 2 GHz by inserting an intra-cavity etalon, although this caused some reduction in power.

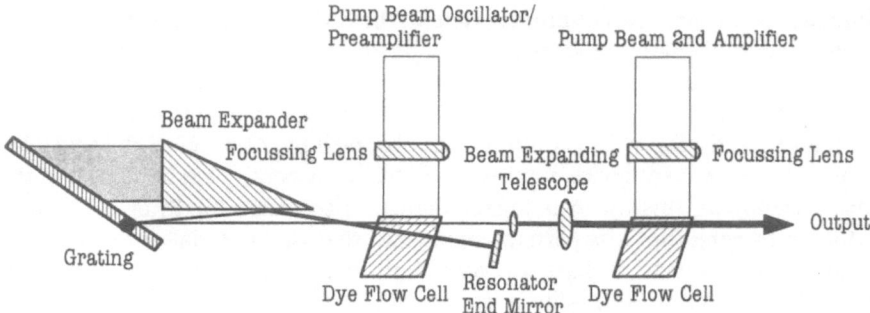

Fig. 5.15. Lambda Physik model FL2001 dye laser: schematic optical layout. Courtesy Lambda Physik GmbH

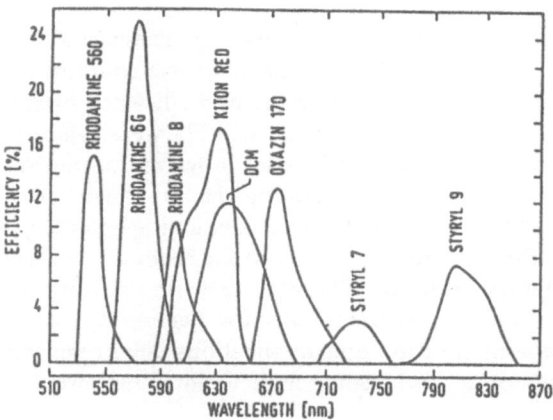

Fig. 5.16. Conversion efficiency and tuning ranges of various laser dyes pumped by a CVL with an unstable resonator (30 W pump power on both lines). The efficiencies are measured for a Lambda Physik FL2001 dye laser without an intra-cavity etalon. From [5.40]

Another commercial dye laser system which can be successfully adapted for CVL pumping is the Lumonics HyperDye 300. The cavity configuration of this laser is shown in Fig. 5.17. Tests carried out by *Evans* [5.41] have shown that, with appropriate choice of output coupler and beam-splitter reflectivity, efficiencies exceeding 28% at the peak of the Rh6G tuning curve can be achieved at bandwidths less than 4 GHz. The frequency stability is better than 1 GHz, and the beam divergence is smaller than 0.5 mrad.

5.5.3 Dye Laser Oscillator-Amplifier Performance

The performance of a CVL pumped dye laser amplifier specially designed to amplify a beam at 627.8 nm from a GVL is described by *Ainsworth* and *Piper*

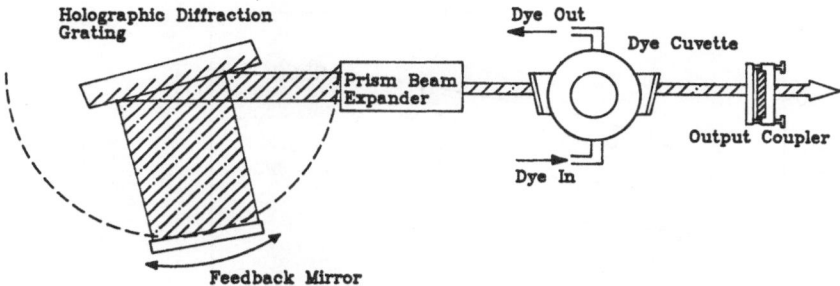

Fig. 5.17. Lumonics dye laser model HyperDYE-300: schematic optical layout. Courtesy Lumonics Inc.

[5.42]. The device is intended for medical applications (Sect. 5.6.1). An interesting feature is that both the CVL beam used for pumping transversely the dye laser amplifier and the GVL beam which forms the input beam for the dye laser amplifier are obtained from the same discharge tube which includes separate hot zones for copper and gold along its length [5.43]. Operating at the fixed wavelength of 627.8 nm on Rh640 in ethanol, efficiencies of 35% were recorded for an injected signal of 50 mW and a pump power of 4 W at 578 nm only. Pumping with both CVL lines decreased the *efficiency* dramatically, but the best output power of about 1 W was obtained with dye mixtures of Rh640 and Rh590 pumped by both CVL lines.

A variable bandwidth CVL-pumped dye laser oscillator-amplifier system in which both the Hänsch-type oscillator and the amplifier dye cell are pumped in a single-sided transverse pumping scheme is described by *Lavi* et al. [5.44]. Various combinations of intra-cavity etalons allow the bandwidth to be chosen within the range 0.4–2.0 GHz. For Rh6G in methanol pumped by 511 nm only, the oscillator efficiency was found to saturate at 15% at pump powers above 1 W, and with both CVL transitions pumping RhB in ethanol oscillator efficiency saturated at 17% even for the maximum bandwidth of 2 GHz. In contrast, the efficiency of the single-stage amplifier is extremely impressive, as shown in Figs. 5.18 and 5.19. At a repetition frequency of 4 kHz, the amplifier efficiency at 605 nm reached 40% on Rh6G for an input of 100 mW, and saturated at 50% for pump power over 3 W. These results are in good agreement with a theoretical model based on that of *Hargrove* and *Kan* [5.20]. Since only a small fraction of the available 20 W of CVL pump power was needed to pump the oscillator, most of the available CVL power could be applied to the amplifier with the result that the overall efficiency of the oscillator-amplifier combination (up to 45%) was only slightly less than that of the amplifier stage alone.

The overall efficiency as well as the overall cost of a high-power dye laser installation involving many stages of CVL-pumped dye laser amplification is

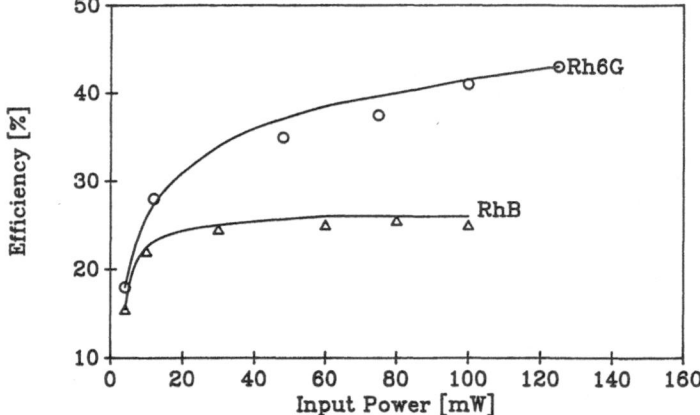

Fig. 5.18. Dye laser amplifier transversely pumped by a CVL: Amplifier efficiency vs. input power from dye laser oscillator. (Experimental data pertains to Rh6G, 0.001 M/l in methanol 580 nm, 2 GHz bandwidth and RhB, 0.025 M/l in ethanol, 605 nm, 2 GHz bandwidth. Pump energy 0.7 mJ per pulse.) From [5.44]

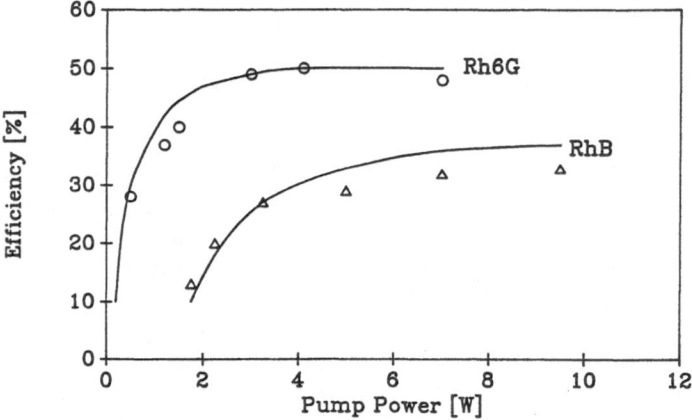

Fig. 5.19. Dye laser amplifier transversely pumped by a CVL: Amplifier efficiency vs. pump power. (Experimental data pertains to Rh6G, 0.001 M/l in methanol, 580 nm, 2 GHz bandwidth, input energy from oscillator 40 μJ per pulse and RhB, 0.025 M/l in ethanol, 605 nm, 2 GHz bandwidth, input energy from oscillator 70 μJ per pulse.) From [5.44]

clearly rather insensitive to the choice of primary oscillator laser. This realization has led to the adoption of Ar^+ laser-pumped cw dye lasers as primary oscillators by some groups involved in AVLIS studies. In particular, the group at the Israel Nuclear Research Center has investigated [5.45] the properties of a CVL-pumped dye laser amplifier system in which the primary oscillator is a Spectra Physics 380D dye laser pumped by a 6 W Ar^+ laser. The experimental

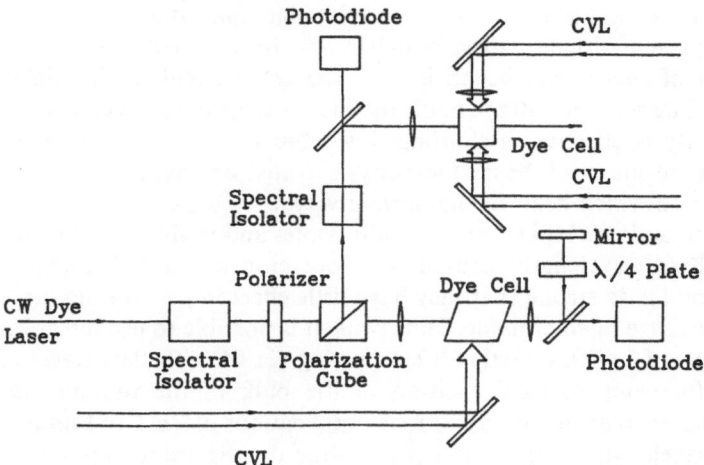

Fig. 5.20. Pulsed high-power dye laser amplifier for cw input beam. Double-pass preamplifier has single-sided transverse CVL pumping. Power amplifier has double-sided transverse CVL pumping. From [5.45]

layout is shown in Fig. 5.20. The plane polarized input beam from the cw dye laser makes a double transit through the transversely pumped pre-amplifier dye cell. Its plane of polarization is rotated through 90° before making the second transit so that, with the aid of a polarization-sensitive beam splitter, the returning amplified beam can be separated from the input beam.

5.6 Applications of CVL Pumped Dye Lasers

5.6.1 Medical

A promising new treatment currently being tested for many forms of cancer is called photodynamic therapy (PDT). This technique makes use of the cytotoxic properties of a light-sensitive compound known as haematoporphyrin derivative (HpD). After receiving an injection of HpD, the patient is required to remain out of strong light for a period while the HpD circulates in the bloodstream, and accumulates in tumours. Some two or three days after injection, the HpD has marked out the regions of cancer, and illumination of these regions with light of the appropriate wavelength causes the HpD to become active in destroying the tumour throughout its volume.

While some details of the cytotoxic process are still imperfectly understood, it is clear that the initial stages [5.46] involve the excitation of the HpD molecule to its first excited singlet level, from which it undergoes a fairly rapid intersystem crossing transition to the lower lying triplet excited state of the molecule. The

HpD molecule can remain in the long-lived triplet state until it collides with a ground state oxygen molecule in the bloodstream. In such collisions, near-resonant transfer of energy can occur, leaving the O_2 molecule in its highly reactive $^1\Delta$ state. The resultant attack made by this active form of oxygen on the tissue in its vicinity is effective in shutting down the supply of blood to the tumour, which in the course of the next few days becomes necrosed and sloughs off to leave a wound which heals in the normal way.

It is possible to activate HpD with both ultraviolet and visible light but the attenuation at these wavelengths caused by absorption in the haemoglobin component of blood is so strong that they have little effect in penetrating more than the first millimetre or so of tissue. Thus while it is possible to use the green and yellow outputs of a CVL directly for eradicating via PDT the last traces of cancerous cells following surgical excision of the bulk of the tumour (for example, in regions such as the brain) for most applications of PDT to tumours of larger mass, wavelengths with greater penetration depths are required.

To avoid the haemoglobin absorption band the wavelength must be longer than 610 nm, but to activate HpD it must be shorter than 640 nm. In early studies of PDT the GVL, with its strong emission at 628 nm, was used extensively.

More recently however, CVL-pumped dye lasers have become the preferred technology. There are several reasons for this. Firstly, with recent improvements in the efficiency of broadband dye lasers, the output power which can now be delivered by a CVL-pumped dye laser is comparable to that obtainable from large gold lasers. Secondly, because the beam quality of the dye laser can be better than that of the GVL, it is easier to match the output into the small diameter optical fibers that are preferred for the delivery system. Thirdly, and perhaps most importantly, is the fact that the output light is tunable in wavelength. This feature is significant because there are a number of efforts worldwide aimed at finding other photosensitizing agents as alternatives to HpD which will offer even higher sensitivity and reproducibility. Among the candidates currently being investigated is aluminium pthalocyanine which responds to radiation in the 680 nm region. Clearly the ability to excite a range of sensitizing agents, each with their own spectral response, is important to medical researchers in this field.

An example of a CVL pumped dye laser system specifically designed for PDT applications is the Oxford Lasers DL20 shown in Fig. 5.21. The compact air-cooled CVL delivers 15 W of pump radiation to the dye laser, which operates at 30% efficiency near the peak of tuning curve of Rh6G. The unit is controlled from a freestanding console, which enables the operator to monitor the power delivered to the fiber and to regulate the light dose via the power meter and shutter built into the dye laser. The delivery system is capable of coupling output simultaneously to four optical fibers of 200–1000 µm diameter and 0.4 numerical aperture which are connected to the disposable treatment tips via standard SMA fittings. The tuning range covers 530–900 nm, and the bandwidth is less than 3 nm.

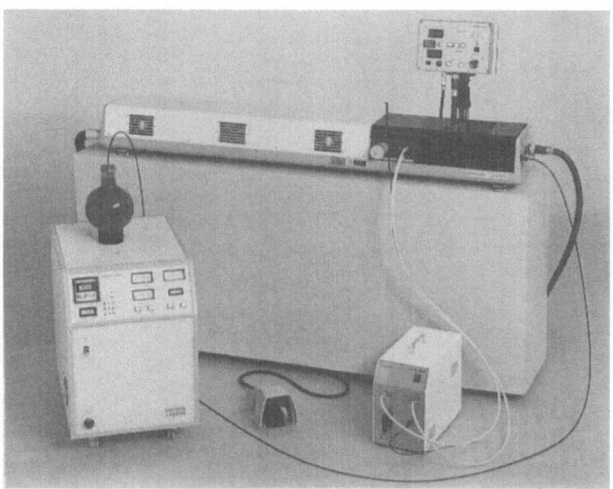

Fig. 5.21. Medical dye laser system for photodynamic therapy. Photograph shows 10 W air-cooled CVL with integral dye laser (Oxford Lasers Model DL20) within optics bay, and control console for dosimetry system. Laser Output via optical fiber leads into flask containing lipid solution for demonstration of PDT in bladder. Courtesy Oxford Lasers Ltd.

5.6.2 Ultrafast Pulse Amplification

Passively mode-locked dye lasers, pumped by cw Ar^+ lasers are a widely used source of subpicosecond pulses. Such a dye laser, operated in the colliding pulse mode-locked (CPM) cavity configuration, is capable [5.47] of generating a continuous train of pulses of 30 fs duration at a repetition frequency of 100 MHz. However, because the pulses are each of very low energy (typically less than 0.5 nJ) their peak intensity is too low for many applications, especially those involving nonlinear phenomena such as "white light" continuum generation. In 1984 a technique for amplifying the femtosecond pulse trains generated by CPM lasers was pioneered by *Knox* et al. [5.48]. The technique has found widespread applications in fields as diverse as solid-state physics and biochemistry.

   As indicated in Fig. 5.22, the optical arrangement involves focusing the pump laser beam onto a dye jet through which the output beam of the CPM laser is passed six times, so that energy is extracted six times from the same gain region of the multi-pass dye laser amplifier by a given femtosecond pulse before it exits. If a low repetition rate device such as a frequency doubled Nd:YAG laser operating at 10–20 Hz is used as the source of pump radiation, despite the large amplification factors that can be achieved, the overall utilization of the 100 MHz femtosecond pulse train is very poor. Only a few pulses in every ten million pass through the amplifier while it has gain.

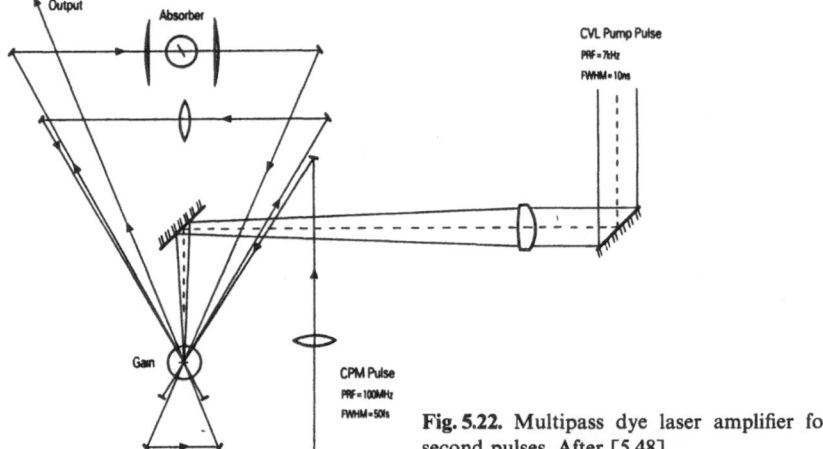

Fig. 5.22. Multipass dye laser amplifier for femto-second pulses. After [5.48]

For this reason the CVL, with its high repetition rate capability, has become the pump laser of choice for this application. However, because the pulses produced by a standard CVL (typically 15–25 ns FWHM) are longer than the 10 ns interval between successive pulses in the 100 MHz pulse train, there is a possibility that two or three neighboring femtosecond pulses could be circulating in the multipass amplifier during a single gain period if such a laser were to be used as a pump. To avoid this problem, CVL cavity designs employing special "short pulse" optics have been devised [5.49] which utilize the abrupt switch from gain to loss which occurs at the end of the CVL's gain pulse to curtail the CVL output pulse and reduce its overall duration to less than 10 ns FWHM. Table 5.2 lists the parameters routinely achieved in day-to-day operation of a femtosecond system employing a multi-pass dye laser amplifier pumped by a 40 W CVL equipped with such a short pulse cavity [5.50].

For systems operating in the picosecond rather than the femtosecond regime, the design of dye cell introduced by *Bethune* [5.51] provides a simpler and cheaper alternative to the free dye jet. A CVL-pumped picosecond dye amplifier system for time-resolved Raman spectroscopy, has been developed at

Table 5.2. Parameters of a Femtosecond Pulse Amplifier

	Output from CPM Laser	Output from Laser Amplifier
Repetition rate	100 MHz	6.5 kHz
Pulse duration	80 fs	100–130 fs
Average power	20 mW	37 mW
Energy/pulse	0.2 nJ	5.75 µJ
Peak power	2.5 kW	57 MW
Photons/pulse	6.2×10^8	1.8×10^{13}

(a)

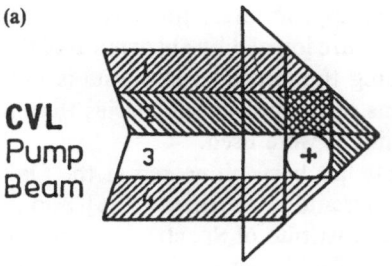

(b)

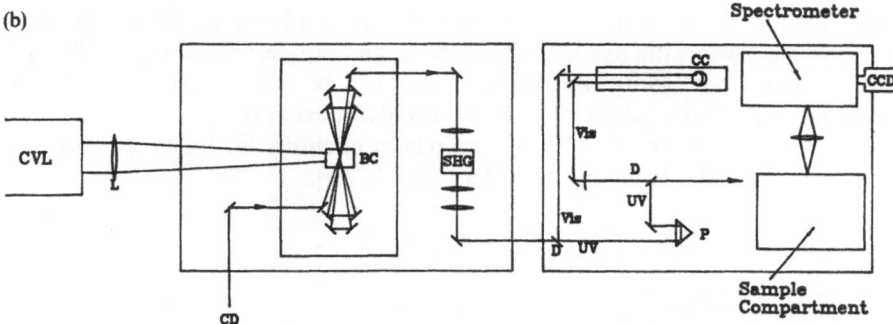

Fig. 5.23 (a) End view of Bethune dye cell prism. (CVL pump beam of 8 mm diameter enters from left. Equal segments 1–4 pump the top, back, front and bottom of the dye solution stream flowing through the 2 mm diameter central channel.) From [5.51]. **(b)** Picosecond Raman spectroscopy system employing CVL-pumped multipass dye laser amplifier with Bethune dye cell. (CD – cavity dumped dye laser input beam; SHG – second harmonic generating crystal; D – dichroic mirror; P – prism; CC – corner cube reflector; CCD – charge coupled device camera.) From [5.52], Courtesy SERC Rutherford Appleton Laboratory

the Rutherford Appleton Laboratory by *Turcu* et al. [5.52]. The system, whose principal features are shown in Fig. 5.23, employs a Bethune cell comprising a right-angled prism of BK7 glass with a 2 mm diameter dye duct running between its parallel faces.

5.6.3 Resonant Ionization Mass Spectrometry – RIMS

An ultra-sensitive technique which allows the quantitative detection of trace quantities of a particular atomic or molecular species present in a sample containing much larger quantities of other species was proposed by *Ambartzumian* and *Letokhov* in 1972 [5.53]. The technique, known as resonant ionization mass spectrometry (RIMS), employs laser radiation tuned to one of the resonance transition wavelengths of the trace species to excite those atoms selectively to one of their resonance levels. The excited atoms can then be photoionized by absorption of laser radiation either directly, or after being promoted to yet higher bound states by stepwise absorption of appropriately tuned laser radiation. Because only atoms of the selected species are ionized by absorption of the tuned radiation, a mass spectrometer sampling the ionized gas plume is presented with a much lower background of ions of the unwanted species than if conventional electron beam ionization techniques were used.

Practical implementation of these concepts has been pioneered at the Oak Ridge National Laboratory in the USA by *Hurst* and co-workers [5.54], and in the USSR by *Letokhov* and co-workers at the Institute of Spectroscopy of the USSR Academy of Sciences, Troitsk [5.55, 56]. The Soviet work, in particular, has employed CVL-pumped dye lasers extensively, indeed the Troitsk group has developed a dye laser design especially for this application. With 8–10 W of CVL pump power, the dye laser operates at an amplifier efficiency of 30% (on Rh6G) and a bandwidth of 24 GHz, which can be reduced to 1 GHz by the addition of an intracavity etalon in the oscillator section.

The reason why the CVL is the pump laser of choice for RIMS applications is evident from the following considerations [5.56].

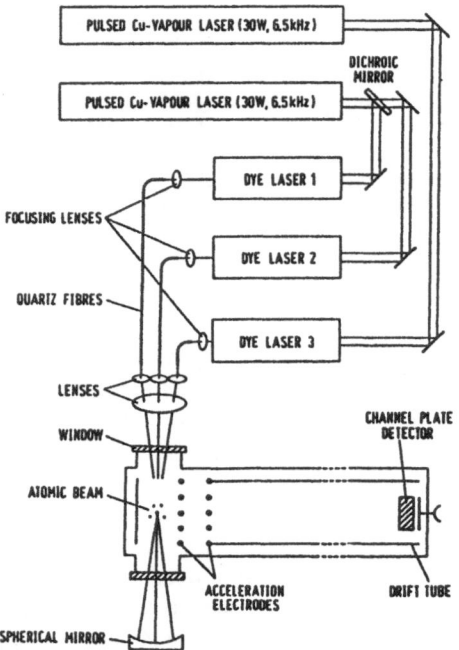

Fig. 5.24. Experimental equipment for the detection of trace quantities of plutonium and technetium by resonance ionization mass spectrometry at the University of Mainz. Three CVL-pumped dye lasers (Lambda Physik Model FL 2001) irradiate the atomic beam at the entrance of a time-of-flight spectrometer. From [5.40]

The average transit time T_{tr} of atoms moving with thermal velocity v_0 through a region of laser irradiation of dimension a is $T_{tr} = a/v_0$. The probability P that any given atom will experience irradiation during its flight through the irradiated region is given by $P = fT_{tr}$ where f is the pulse repetition frequency. In order that all of the atoms should have a good probability of being irradiated (i.e. that P should be of order unity) f must be of order $1/T_{tr}$.

In a typical RIMS apparatus, the laser beam diameter is of order 1–10 mm, and thermal velocities are typically of order $100 \mathrm{\,m\,s^{-1}}$. To satisfy the requirement that every atom passing through the irradiation zone should have a chance of experiencing irradiation requires that the pulse repetition frequency f should be of order 10–100 kHz. Among currently available pulsed laser systems, the CVL uniquely has the ability to satisfy this requirement with an appreciable energy per pulse.

To establish the sensitivity which can be achieved using the RIMS technique to detect small quantities of plutonium and technetium, experiments [5.40] have been carried out by *Kluge, Trautmann* and co-workers at the University of Mainz, Germany. An atomic beam was formed by evaporation of the sample material from an incandescent rhenium filament impregnated with a fine layer of the material. As shown in Fig. 5.24, at the entrance to the time-of-flight spectrometer, the atomic beam was irradiated by the beams from three dye lasers (Lambda Physik FL2001) pumped by two CVLs (Oxford Lasers Cu40) fitted with unstable resonators. The dyes used covered the spectral range from 530–840 nm. The conversion efficiency exceeded 10% in most cases and reached 25% in the case of Rh6G (Fig. 5.16). The dye laser bandwidth was measured as 3.5–10 GHz depending on the dye chosen, but this could be reduced to 0.8 GHz by the insertion of an intracavity etalon [5.40].

For the excitation of plutonium atoms several RIS schemes were investigated, but all used a wavelength of 586.5 nm to promote atoms from the $5f^6 7s^2$ $^7F_0$ ground state to the $5f^6 7s7p\ ^9G_1$ resonance level as the first excitation step. The wavelength of the second step was selected from the several available transitions in the range from 655–688 nm. For the final photoionizing step various autoionizing transitions in the range from 570–580 nm were tested.

In the case of technetium it was necessary to use frequency-doubled radiation in the range 313–318 nm to excite atoms from the $4d^5 5s^2\ ^6S_{5/2}$ ground state. In the second step, levels of the $4d^6 6s\ ^4D$ or ^6D terms were excited by absorption of infrared radiation in the range 787–834 nm. From these states, autoionizing levels could be reached by absorption of radiation in the range 560–690 nm.

Defining the detection efficiency as the ratio of ions counted in resonance to the number of atoms evaporated, the Mainz group calculate that the efficiency is 2.4×10^{-6} for the detection of plutonium atoms, implying a detection limit of 2×10^6 atoms per isotope. The corresponding limit for the detection of technetium, again based on extrapolation from samples containing 10^{12}–10^8 atoms of this element, was found to be 4×10^6 atoms per sample.

5.6.4 Atomic Vapor Laser Isotope Separation (AVLIS)

The atomic vapor laser isotope separation (AVLIS) process for enriching the content of ^{235}U in natural uranium makes use of the fact that many of the visible transitions of ^{238}U display a considerable frequency shift (several times the Doppler width) from the corresponding transitions of ^{235}U. It is thus possible, by irradiating a stream of uranium vapor evaporated by electron beam heating a crucible containing the solid metal, to ionize selectively the ^{235}U atoms by stepwise absorption of radiation from three different dye laser beams tuned to the frequencies which lead to excitation and finally ionization of ^{235}U. Having passed through the irradiation zone, the vapor stream contains ions of ^{235}U which are deflected by an electric field and deposit as solid material on the charged plates. The atoms of ^{238}U pass through undeflected to be collected as the tails of the separation process on another set of plates.

Apart from the extra refinement of restricting the frequency spread of the dye laser beams to cover the absorption lines of one isotope only, the laser requirements of the AVLIS process are analogous to those of the RIMS technique described in Sect. 5.6.3. Just as in RIMS, the requirement is for a pulsed laser of high-repetition frequency to ensure that each atom traversing the irradiation zone has a high probability of experiencing irradiation. The CVL is therefore the pump laser of choice for this application too.

The US program of CVL development for AVLIS applications is well documented in the series of annual reports from LLNL up to 1981, at which point it seems that CVL amplifier units of up to 200 W per unit were being developed. As the work has reached commercial viability fewer details of the technology have been released for publication. However, *Paisner* [5.8] reports that by 1985 there were installed in Livermore six chains of CVL amplifiers, each

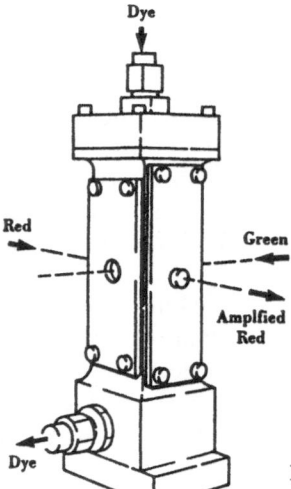

Fig. 5.25. High-power dye laser amplifier for AVLIS studies. Courtesy AEA Technology

containing 30 laser units with a total output capacity of several thousand watts. This report further indicates that the largest dye laser system in the world now exists at LLNL with more than 900 W average power available in multicolor dye laser beams and 600 W in a single-color dye laser beam at repetition rates in excess of 4 kHz. The laser systems are currently being commercialized for industrial-scale processing of reactor fuel by Martin Marietta Energy Systems in Oak Ridge, Tennessee.

At Capenhurst in the UK, British Nuclear Fuels plc is also exploring the commercial viability of AVLIS using an array of 100 W CVL lasers (Oxford Lasers Cu100) to pump high power dye laser amplifiers of the type shown in Fig. 5.25 developed by AEA Technology at Culham Laboratory.

References

5.1   G.R. Fowles, W.T. Silfvast: Appl. Phys. Lett. **6**, 236 (1965)
5.2   W.T. Walter, M. Piltch, N. Solimene, G. Gould: Bull. Am. Phys. Soc. **11**, 113 (1966)
5.3   A.A. Isaev, M.A. Kazaryan, G.G. Petrash: ZhETF Pis. Red. **16**, 40 (1972) [English transl.: JETP Lett. **16**, 27 (1972)]
5.4   A.A. Isaev, G.Yu. Lemmerman: Kvant. Elektron. **4**, 1413 (1977) [English transl.: Sov. JQE **7**, 799 (1977)]
5.5   R.E. Grove: Copper vapor laser comes of age, in Laser Focus, July 1982, pp. 45
5.6   I. Smilanski, A. Kerman, L.A. Levin, G. Erez: Opt. Commun. **25**, 79 (1978)
5.7   R.S. Anderson, B.E. Warner, C. Larson Sr., R.E. Grove: CLEO Conf. 10–12 June 1981, Dig. of Tech. Papers (Opt. Soc. Am. 1981)
5.8   J.A. Paisner: Appl. Phys. B **46**, 253 (1988)
5.9   M.J. Kushner, B.E. Warner: J. Appl. Phys. **54**, 2970 (1983)
5.10 R.R. Lewis: Mechanisms of copper vapour lasers, D.Phil. Thesis, University of Oxford (1985)
5.11 C.S. Liu, E.W. Sucov, L.A. Weaver: Appl. Phys. Lett. **23**, 92 (1973)
5.12 C.J. Chen, N.M. Nerheim, G.R. Russell: Appl. Phys. Lett. **23**, 514 (1973)
5.13 D.N. Astadjov, N.K. Vuchkov, N.V. Sabotinov: IEEE J. Quantum Electron. **24**, 1927 (1988)
5.14 Proc. Lebedev Physics Institute, Acad. of Sci. USSR, Vol 181 "Metal Vapor and Metal Halide Vapor Lasers", ed. G.G. Petrash (Nova, New York 1989)
5.15 K.I. Zemskov, A.A. Isaev, M.A. Karazyan, G.C. Petrash, S.G. Rautian: Kvant. Elektron. **1**, 863 (1974) [English transl.: Sov. JQE **4**, 474 (1974)]
5.16 A.A. Isaev, M.A. Kazaryan, G.C. Petrash, S.G. Rautian, A.M. Shalagin: Kvant. Electron. **4**, 1325 (1977) [English transl.: Sov. JQE **7**, 746 (1977)]
5.17 G.A. Naylor, A.J. Kearsley, R.R. Lewis: Oxford Lasers Technical Note: No. 1 "Cavity designs for Metal Vapor Lasers"; Oxford Lasers Ltd. (1988)
5.18 R.R. Lewis, G.A. Naylor, A.J. Kearsley: Copper vapor lasers reach high power, in Laser Focus, pp. 92–96, April 1988
5.19 M. Amit, S. Lavi, G. Erez, E. Miron: Opt. Commun. **62**, 110 (1987)
5.20 R.S. Hargrove, T. Kan: IEEE J. Quantum Electron. **16**, 1108 (1980)
5.21 O. Teschke, A. Dienes, J.R. Whinnery: IEEE J. Quantum Electron. **12**, 383 (1976)
5.22 M. Amit, G. Bialolenker, D. Levron, Z. Burshtein: J. Appl. Phys. **63**, 1293 (1988)
5.23 W. Sun, C.S. Tang, X.B. Zhuge, M.S. Chen: Opt. Commun. **58**, 196 (1986)
5.24 P.R. Hammond: Fire suppression in laser dye media, in LLNL Laser Program Annual Report 1980 (UCRL-50021-80) pp. 10–27, 10–28
5.25 D.W. Phillion, D.J. Kuizenga, A.E. Seigman: J. Chem. Phys. **61**, 3828 (1974)
5.26 A. Wokaun, P.F. Liao, R.R. Freeman, R.H. Storz: Opt. Lett. **7**, 13 (1982)
5.27 T.J. Chuang, K.B. Eisenthal: Chem. Phys. Lett. **11**, 368 (1971)

5.28 R.S. Hargrove, T. Kan, A.A. Pease: Dye amplifiers, Sect. 8–6, in LLNL Laser Program Annual Report 1976, pp. 8–36, 8–42

5.29 W.W. Morey: In Proceedings of the International Conference on Lasers '79, ed. by V.J. Corcoran (STS, McLean, VA 1980) pp. 365–373

5.30 Z-G. Huang, K. Namba: Jpn. J. Appl. Phys. **20**, 2383 (1981)

5.31 A.A. Pease, W.M. Pearson: Appl. Opt. **16**, 57 (1977)

5.32 T.W. Hänsch: Appl. Opt. **11**, 895 (1972)

5.33 F.J. Duarte, J.A. Piper: Appl. Opt. **21**, 2782 (1982)

5.34 F.J. Duarte, J.A. Piper: Appl. Opt. **23**, 1391 (1984)

5.35 F.J. Duarte: Narrow-linewidth pulsed dye laser oscillators, in Dye Laser Principles, ed. by F.J. Duarte, L.W. Hillman (Academic, New York 1990) pp. 133–183

5.36 A.F. Bernhardt, P. Rasmussen: Appl. Phys. **B26**, 121 (1981)

5.37 A.N. Zherikin, V.S. Letokhov, V.I. Mishin, V.P. Belyaev, A.N. Evtyunin, M.A. Lesnoi: Kvant. Elektron. **8**, 1340 (1981) [English transl.: Sov. JQE **11**, 806 (1981)]

5.38 M. Broyer, J. Chevaleyre, G. Delcretaz, L. Woste: Appl. Phys. **B35**, 31 (1984)

5.39 K.L. Hohla: Excimer-pumped dye lasers – the new generation, in Laser Focus, pp. 67–74, June 1982

5.40 W. Ruster, F. Ames, H.J. Kluge, E-W. Otten, D. Rehklau, F. Scheerer, G. Hermann, C. Mühleck, J. Riegel, H. Rimke, P. Sattleberger, N. Trautmann: Nucl. Instrum. Meth. A**281**, 547 (1989)

5.41 I.J. Evans: Clarendon Laboratory, Oxford. Unpublished data (1990)

5.42 M.D. Ainsworth, J.A. Piper: Opt. Commun. **69**, 294 (1989)

5.43 A.J. Kearsley, R.R. Lewis: CLEO Conf. 21–28 May 1988, Dig. Tech. Papers (Opt. Soc. Am. 1988)

5.44 S. Lavi, M. Amit, G. Biololanker, E. Miron, L.A. Levin: Appl. Opt. **24**, 1905 (1985)

5.45 S. Lavi, G. Biololanker, M. Amit, D. Belker, G. Erez, E. Miron: Opt. Commun. **60**, 309 (1986)

5.46 J-L. Boulnois: Lasers Med. Sci. **1**, 47 (1986)

5.47 J.A. Valdmanis, R.L. Fork, J.P. Gordon: Opt. Lett. **10**, 131 (1985).

5.48 W.H. Knox, M.C. Downer, R.L. Fork, C.V. Shank: Opt. Lett. **9**, 552 (1984)

5.49 Oxford Lasers Technical Note: No. 3 "Short Pulse Generation for Metal Vapor Lasers"; Oxford Lasers Ltd. (1988)

5.50 J.F. Ryan, R.A. Taylor: private communication

5.51 D.S. Bethune: Appl. Optics **20**, 1897 (1981)

5.52 I.C.E. Turcu, J. Diggins, M. Hannah, F. O'Neill, I.N. Ross, S. Bell, R.E. Hester, S. Umapathy, J.M. Barr: Picosecond pulse dye amplifier pumped by a 6.7 kHz copper laser, Annual Report to the Laser Facility Committee 1989, SERC Rutherford Appleton Laboratory, RAL-89-045, pp. 382–385 (1989) Annual Report to the Laser Facility Committee 1990, SERC Rutherford Appleton Laboratory, RAL-90-026, pp. 185–186 (1990)

5.53 R.V. Ambartzumian, V.S. Letokhov: Appl. Opt. **11**, 354 (1972)

5.54 G.S. Hurst, M.G. Payne: Principles and Applications of Resonance Ionization Spectroscopy (Adam Hilger, Bristol 1988)

5.55 V.S. Letokhov (ed.): Laser Analytical Spectrochemistry (Adam Hilger, Bristol 1986)

5.56 V.S. Letokhov: Laser Photoionization Spectroscopy (Academic Press, Orlando 1987)

6. Flashlamp-Excited Dye Lasers

Patrick N. Everett

With 8 Figures

Flashlamp-excited dye lasers (FEDLs) are versatile sources of coherent light. A single dye will typically tune through 50 nm or more. The same hardware can be used, with alternative dyes, to cover a spectrum of about 300 nm. Spectral bandwidths range from a few nanometers down to several femtometers ($\cong$ 10 MHz), in pulses of picoseconds to milliseconds. The basic technology is relatively simple, and many experimenters have built FEDLs. The medium is inexpensive and easy to renew. It has high heat capacity, and the active volume can be replaced between pulses. As much as 400 J/pulse, and 1.2 kW average power in a burst mode, have been obtained, and much potential remains.

Applications include isotope separation, photochemistry, excited state and non-linear spectroscopy, underwater illumination and communication, medicine, atmospheric sounding, pollution detection, plasma diagnostics, and research in physics, chemistry and biology. Overall efficiency rarely exceeds 1.5%, but improvement is expected as better dyes become available and spectral transfer taps more of the flashlamp energy. Replenishment of dye while the laser is running resolves the dye life problem. Diffraction limited beams are available, but at some sacrifice in energy. Thermal effects and triplet accumulation can reduce beam quality and efficiency for pulses longer than about 1 μs. Flashlamps are fragile, but with good technology and correct mounting and operation, each flashlamp can last many millions of pulses. They can be driven harder, consistent with useful life, than previously thought.

This review is aimed toward the "hands on" practitioner. It will start with a brief history of key developments in the field. It will then proceed through an overview of design considerations into more detailed discussion of the technology and useful modeling to aid insight and design ability. The work complements earlier reviews [6.1–8].

6.1 Development of FEDLs

The first successful FEDL was demonstrated by *Sorokin* and *Lankard* in 1967 [6.9], with a coaxial flashlamp containing argon and air. They obtained lasing with four dyes in the mid-visible. The strongest was rhodamine 6G, which still

Springer Series in Optical Sciences, Vol. 65
High-Power Dye Lasers Editor: Francisco J. Duarte
© Springer-Verlag Berlin Heidelberg 1991

remains the standard of comparison. Further detail on these early results was published by *Sorokin* et al. in 1968 [6.10].

Soon after, *Schäfer's* team [6.1] successfully used commercial linear xenon-filled flashlamps, and for some dyes even helical flashlamps, provided the rise time was as short as about one microsecond to avoid triplet accumulation. *Snavely* and *Schäfer* [6.11] found that oxygen in the solution could quench triplets, obtaining pulses lasting 140 μs when pumping rhodamine 6G in methanol with a 500 μs pulse. They attributed the early termination to thermal and acoustic effects rather than to triplet accumulation. These results were extended by *Marling* et al. [6.12, 13], and *Pappalardo* et al. [6.14], who used unsaturated hydrocarbons and oxygen for triplet quenching. They obtained pulse durations of 600 μs using COT as a triplet quencher, and argued that triplets were more important than thermal problems in limiting pulse length. In 1973, *Ewanizky* et al. [6.15] demonstrated that lasing could be terminated by shock waves from a coaxial lamp discharge. In 1974, *Blit* et al. [6.16] showed that infrared heating of the medium could terminate lasing. *Janes* et al. demonstrated a novel approach in 1987 for overcoming all these effects [6.17]. It synchronously sweeps one or more FEDL pump beams and the axis of the resonator across the face of the laser dye cell at supersonic speed, leaving thermal distortions and triplet accumulations behind.

Passive mode-locking of FEDLs was achieved by *Schmidt* and *Schäfer* in 1968 [6.18], with pulses shorter than 0.4 ns. The early work is summarized by *Shank* and *Ippen* [6.19]. In 1987 *Singh* [6.20] reported pulse lengths of less than 1.5 ps. *Morton* et al. in 1978 [6.21] showed that intermediate pulse lengths of 20 ns can be obtained by cavity dumping.

In 1963, *Emmett* and *Schawlow* [6.22] found peak output and life of linear lamps can be increased by having a low-level discharge prior to the main discharge. This was applied to FEDLs in 1974 by "simmering" with a low-level dc discharge between pulses [6.23], and "pre-pulsing" just prior to the main pulse [6.24]. In 1976 *Friedman* and *Morton* [6.25] obtained 0.3 J/pulse at 30 Hz with pre-pulsed lamps. This laser also introduced transverse dye flow through the laser head to give better clearing of the dye at the higher pulse rate. In 1978 *Jethwa* et al. [6.26] achieved more than 2 J/pulse using four simmered linear xenon filled lamps.

Recent high-power FEDLs have generally used linear flashlamps filled with xenon and simmered or pre-pulsed to increase output and life. Coaxial lamps are suitable for lasers with pulse rates below about 10 Hz. Air-filled ablating-wall flashlamps [6.27] have been used. In 1972 *Gibson* [6.28] obtained 10 W average power from an FEDL using such lamps. In the same year *Anliker* et al. [6.29] used eight ablating-wall lamps to obtain 12 J, 5 μs pulses, with overall efficiency of 1.2%. Vortex stabilized lamps were used in 1976 by *Mack* et al. [6.30] to obtain 90 W of average power at 0.26% efficiency, and by *Morey* and *Glenn* [6.31] to obtain 114 W average power at 255 Hz. Continuous wave operation, but at low output, was obtained by *Thiel* et al. [6.32] in 1987, pumping rhodamine 6G with modified high-pressure arc lamps.

As much as 400 J has been obtained from an FEDL [6.33], pumping rhodamine 6G with a coaxial lamp with 50 kJ input (0.8% efficiency). A commercial laser has delivered more than 100 J in 2 µs pulses using a coaxial flashlamp [6.7]. Average power of 200 W at 50 Hz and 0.6% efficiency has been obtained from a transverse dye flow planar waveguide laser [6.34], using simmered linear flashlamps, COT as a triplet quencher, and a spectrally converting dye solution between the flashlamps and reflectors. In a burst mode, 1.2 kW has been reported [6.35] using a transverse flow FEDL with simmered linear lamps.

Efficiency as high as 1.8% has been published [6.36] and 2% has been reported privately [6.37]. Various experimenters have explored shifting wasted lamp radiation to wavelengths useful for pumping the lasing dye with an intermediate dye [6.38–43].

Laser dyes are commercially available covering the spectrum from 340 nm to 1 µm. This range can be extended down to 250 nm by frequency doubling [6.44, 45]. Over 400 dyes have lased, many of them specially synthesized, and the number steadily grows. The dye degradation problem is being overcome by real time dye cleaning.

A spectral width of 2.5 GHz, in 0.3 J pulses, was obtained by *Gibson* in 1972 [6.28] using tilted solid etalons. It was used to excite sodium in the upper atmosphere. *Gale* [6.46] obtained a spectral width of 4 MHz in 1973, stable within 12 MHz, with 2.5 mJ/pulse, by tuning an FEDL oscillator with three air-spaced Fabry-Pérot etalons. Spectral width controllable down to 0.5 GHz, at 0.1 J/pulse, was obtained in 1986 by the author and co-workers using an FEDL oscillator and amplifier with a waveguide configuration in one dimension. The oscillator had up to three solid uncoated etalons, and the output was coupled with a diffraction grating at grazing incidence. It had intermode spacing of 50 MHz, pulse length of 4 µs, and exceeded 10 Hz.

Narrower spectra, with higher energy, have been obtained from oscillators injection-locked by single-mode cw dye lasers (laser excited). Gains of more than 10^6 have been achieved with the efficiency of a saturated oscillator. *Magyar* and *Schneider-Muntau* in 1972 [6.47] obtained 0.6 J in 8 GHz bandwidth (0.01 nm). *Maeda* et al. in 1975 [6.48] obtained 4 J in 4 GHz (0.005 nm), with efficiency 0.24%. Two years later, *Blit* et al. [6.49] obtained 50 mJ pulses in 30 MHz. *Boquillon* et al. in 1987 [6.50] obtained 4 mJ in 6 MHz, in 0.4 µs pulses at 10 Hz.

High-pulse-rate injection-locked ring oscillators, using transverse-dye-flow FEDLs, were developed at Exxon Nuclear Corporation and Avco Everett Research Laboratory (now Avco Research Laboratory, Inc.) between 1970 and 1980 for enriching uranium from its atomic vapor. Pulse spectra were tailored in the sub-GHz regime for selectively exciting ^{235}U through successive levels to ionization. Techniques were developed for filling mode gaps and for chirping mode structure for matching the uranium spectra [6.51]. The work reached the system prototype level. Although economically attractive, the project was eventually abandoned because the U.S. Administration of that period turned to discouraging private enriching of uranium. The work was done under a cloak of proprietary secrecy, but some results have crept into general practice as a result of personnel mobility. An idea of the advances achieved can be gleaned from some of the patents issued during those years [6.51–72].

A transportable laser for rugged use is being developed to deliver more than 100 W average power at 40 Hz for underwater communications by Avco Research Laboratory, Inc. [6.73]. It uses coumarin 450 dye, with a filter and a dye replenishment system. It includes Oligo 373 dye as a converter in a separate channel, and a planar cavity geometry to maximize the effect of the converter.

A six-beam system, to deliver 300 W average power, 5 J/pulse, is being assembled by MIT Lincoln Laboratory at a remote site for experiments [6.74]. The laser heads were developed by Candela Laser Corporation, and the power supplies by A.L.E. Systems Inc. It uses coumarin 504 dye in aqueous acetamide solvent, with oxygen bubbled into the solution for triplet quenching, and includes a real-time dye regeneration system. The solvent is not flammable, and so eliminates the fire hazard usually associated with the quantity of solvent needed in such large systems.

Standard commercially obtainable lasers include the Candela Laser Corp model LFDL-40, that gives 10 J/pulse at 10 Hz using pyrromethene dye with longitudinal flow, the TFDL-50 that gives 1 J/pulse at 100 Hz with transverse flow, and the Phase-R Corporation # 9102 that gives 100 J/pulse at 4 ppm. The latter company has also done a paper study [6.75] showing feasibility of producing 1 kJ/pulse at 1 Hz.

6.2 General Design Considerations

Some of the major design considerations will be summarized in this section, but the more detailed analysis will be deferred to later sections. Dye lasers are typically "4-level" systems. Molecules are excited from low-lying vibrational levels of the ground electronic state, into vibrationally excited levels of the first or second excited singlet states. Within picoseconds they relax non-radiatively into the lower vibrational levels within the same state, which form the upper level of the lasing transition. They can then be stimulated to relax radiatively back into the vibrational complex of the ground state, or they decay spontaneously with a lifetime usually measured in nanoseconds. From the termination of this lasing transition, the molecules relax non-radiatively, within picoseconds, to the lower vibrational levels where they started. Dye molecules are large and have many vibrational possibilities, so bandwidths of dye lasers are quite broad, typically about 50 nm. This allows the dyes to be tuned over wide spectral ranges, and to be excited by the broad spectrum from flashlamps. Because of the rotational complex of each vibrational level, the spectrum is quasi-continuous.

For efficiency, the lasing intensity must exceed I_{sat}, the lasing intensity just sufficient to cause 50% of the emissions to be stimulated rather than spontaneous. This is also the intensity which saturates the exponential laser gain to

50% of the small signal value. Hence an efficient laser is a saturated one; and oscillators, with their inherently saturated operation, are more efficient than high-gain amplifiers.

Dye efficiency increases so fast with the pump intensity that normally the lamps are run as hot as is consistent with life requirement. Color matching would be better if driven less hard, but the other effects are more important.

6.2.1 Where Does the Energy Go?

Efficiency from lamp input to lasing output can be as high as 2%, but 0.25%–1% is more typical. Only a small portion of the light output from the flashlamp is within the absorption band of the dye, often as low as 2%. Dye concentration can be increased to absorb in the weak wings, but this causes uneven heat deposition which degrades the beam quality. A more promising approach is to spectrally shift some of the wasted light to the useful band with an intermediate dye.

For a practical example, we will trace the energy flow through an actual system [6.74] being deployed at an experimental field site. Each of its six output beams comes from an oscillator containing two laser heads. Each head is excited by two flashlamps, 61 cm long and 7 mm bore. The four lamps are excited with a total of 2 kJ for each pulse, and about 5 J of lasing output is obtained, giving an overall efficiency of 0.25%. For the six beams, at 10 Hz, the lamp inputs total 120 kW. Further power is needed for pumping the fluids, for lamp simmer and for other services. Because of the high powers, the operating efficiency in such a laser is an important parameter. Table 6.1 summarizes the contributing efficiencies and indicates where the energy goes, mostly into the flashlamp coolant. Lesser portions go into the laser housing and into heating the dye solution. It is among the designer's goals to obtain a satisfactory portion in the laser output, with the medium sufficiently unperturbed to allow acceptable beam quality.

The flashlamps contain some of the key technology and drive the sizes of the laser heads and the thermal and power budgets. They provide a good starting point for more detailed discussion.

Table 6.1. Modeled efficiency balance for a 300 W average power FEDL system

Step	Modeled efficiency	Lost power
Electric power to flashlamp output within pumping band	2.9%	116.5 kW
Coupling to pass into dye	50%	1.74 kW
Absorbed by dye	57%	760 W
Converted by dye into lasing output	30%	700 W
Overall efficiency and lost power	0.25%	119.7 kW

6.3 Flashlamps

The surface intensity obtainable from flashlamps, consistent with life, generally limits the output and efficiency of the laser system. Flashlamps were originally devised for high speed photography. They have been used in various shapes, including linear, pi, helical, coaxial, triaxial and quadraxial. *Marshak* [6.76, 77], *Goncz* [6.78] and *Furumoto* and *Ceccon* [6.79] gave useful early reviews of the technology and many practical details. More recent overviews are by *Furumoto* [6.6] and in the publications [6.80–82]. Other major manufacturers also provide helpful bulletins.

High-power applications generally use linear flashlamps, with a tubular quartz envelope about 1 mm thick, internal bore measured in mm, and length ranging from a few to many cms. Tungsten electrodes are sealed into the ends, and the tube filled with a noble gas, commonly Xe because it is more efficient, but occasionally Kr. Mixed gases give lower conversion efficiency [6.82]. The tip of the cathode is generally a tungsten matrix impregnated with an emissive material that decreases the work function. The tip of the anode is usually tungsten, sometimes thoriated. In some cases it is the same as the cathode to limit damage from current reversal.

More than 90% of the input power becomes heat, close to the lamps. Water cooling is common. Often a transparent tube surrounds the lamp to carry the coolant. It may be of pyrex glass to absorb UV light that will cause thermal distortion and rapid degradation of the dye. The coolant may contain a UV absorber, or a spectral-converter to shift unused light to a wavelength that can pump the dye. Sometimes the lamp and the dye cell are immersed in flowing coolant filling the pumping cavity.

The lamps are excited by charging a capacitor bank to between 5 kV and 40 kV and then discharging the bank into the lamps through a series switch and short cables of low inductance. The pulse length is determined mainly by the storage capacitance and the inductance of the circuit including the lamps. Relatively short lamp pulses, usually less than 10 µs, are used. This is to minimize triplet accumulation in the dye (discussed in Sect. 6.5), permit high instantaneous power density and minimize thermal-acoustic distortions which develop at the speed of sound in the medium (about 1 mm/µs). High voltage allows sufficient energy to be stored in a capacitance small enough to be consistent with the short pulse required.

For the majority of flashlamp applications, the plasma temperature is in the range 9000–12 000 K during the flash, but the pumping requirements of typical dye lasers require hotter lamps in the range of 20 000 K. This results in stronger emission at the shorter wavelengths and makes a black body approximation more realistic. The intense UV at wavelengths below 224 nm can produce ozone from air. This is toxic and can destroy O-rings and plastic components if not filtered in some manner, as by a pyrex jacket containing cooling water.

The first FEDL [6.9] used a coaxial lamp in which a central tube contains flowing dye and the discharge is created in the annular surrounding space. Many lasers still use coaxial lamps, particularly for pulse rates lower than a few Hz. The triaxial tube adds an annular space between the dye tube and the plasma to carry cooling water [6.3], and reduces the thermal distortions in the dye. The quadraxial tube adds a further annulus, between the dye and the cooling fluid, to carry cooling gas and better thermally isolate the dye [6.83]. *Hirth* et al. [6.84] compared coaxial and linear lamps and concluded that higher efficiency and beam quality can be expected with linear lamps.

Two types of seal construction are suited to the high excitations required [6.82]. The "rod" or graded seal construction has the advantage that it can be processed at the higher temperatures desirable for ensuring maximum reliability and life. The "end cap" or solder seal construction is more rugged, but the low melting point (350°C) of the lead-indium solder is a disadvantage.

6.3.1 Simmer, Lamp Efficiency and Life

The lamp discharge begins as a spark filament of ionized gas between the electrodes. With no pre-ionization, the initial filament forms along the inside wall of the lamp, and then grows with the appearance of a shock wave until it fills the bore, absorbing energy. Before dye lasers, it was found that pre-pulsing enhances UV emission [6.22], and lamp efficiency is higher after the discharge fills the tube [6.85]. After the discharge fills the bore, more energy goes into heating the plasma and less into the shock wave [6.86]. In 1974, it was demonstrated that about 30 mA of "simmer" between the pulses [6.23] improves life and reduces jitter, and pre-pulsing reduces the main pulse's rise time and increases efficiency [6.24]. In 1976 *Friedman* and *Morton* [6.25] introduced 20 A pulses, 0.5 μs before the main pulse, for the same purpose.

When the lamp is simmered or pre-pulsed, the discharge has less contact with the walls, giving more consistent output and longer life. The pulse tends to shorten because the initial breakdown delay is avoided. The shorter pulse leads to higher intensity, higher lasing efficiency and less dye triplet accumulation. In 1976, simmer and pre-pulsing were combined to obtain yet higher efficiency and reproducibility [6.87]. In 1979, *Yee* et al. [6.88] were successful with 30 mA cw simmer, raised to 3 A about 100 μs before the main pulse. Their publication includes useful circuit diagrams. Continuous simmer of at least 2 A gave the same improvement but at a cost in power. At higher rep-rates the cost of the continuous 2–3 A simmer is less significant, and the system is simpler. In 1980 *Basov* et al. [6.89] published further data on simmering.

Flashlamp failure is generally in one of two modes [6.78]. When the pulsed input approaches the "ultimate energy," the lamp will usually fail catastrophically within a few hundred pulses. For lesser energies, the lamp output slowly deteriorates with repeated pulses, from erosion of the electrodes and walls and a gradual buildup of deposits within the envelope. Millions of flashes can be obtained if the excitation energy is reduced towards about 10% of the ultimate energy, provided that the best technology is used in making the lamps.

6.3.2 Pulse Energy, Shape and Length

The energy stored in the capacitor bank is $CV^2/2$, where V is the voltage and C is the bank capacitance. The lamp resistance depends upon the discharge current, and for most of the pulse is given by [6.78]

$$R = 1.13L_A(IA_C)^{-1/2} , \tag{6.1}$$

where L_A is the arc length (cm), A_C is the cross sectional area (cm^2), and I is the instantaneous current (amps). The $I^{-1/2}$ dependence is discussed by *Markiewicz* and *Emmett* [6.90]. It is applicable if the discharge is in Xe plasma filling the bore. If the discharge is very intense, then silica ablating from the wall may reduce the validity later in the pulse. The instantaneous power input is I^2R.

Markiewicz and *Emmett* [6.90] investigated the non-linear oscillatory equations that result from the discharge into the inherent inductance L of the circuit including the lamp. They normalized time to the constant $\tau = \sqrt{LC}$, and plotted computer solutions for the development of current and power input for different values of a "damping term" α determined by the electrical parameters. The highest peak power input is obtained with $\alpha \approx 0.8$, corresponding to critical damping (i.e., just no overshoot of the current). This is generally considered optimum, with sometimes a little overshoot to help quench the switching device after the pulse is over.

It will be shown later that, to a good approximation, a lamp can be considered a black body in quasi-equilibrium, with the radiating power balancing the input I^2R power. The temperature is determined by the input power per

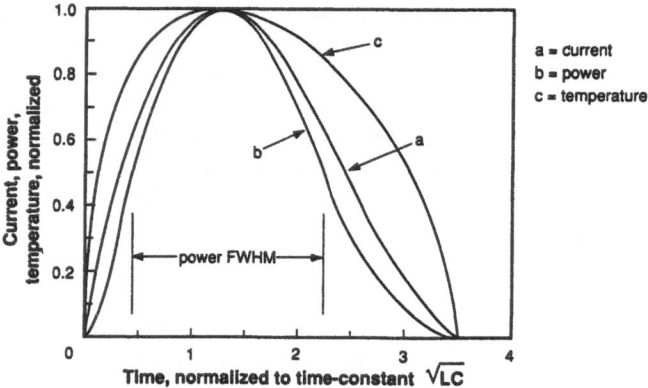

Fig. 6.1. Time-development of current, power and temperature in a flashlamp with $\alpha = 0.8$, (*a*) and (*b*) current and power (from [6.90]), (*c*) temperature from black body model

Table 6.2. Duration of critically damped pulses in terms of time constant $\tau = \sqrt{LC}$

	Rise time 10%–90%	Full width Half Max.	Between 1/3 points	Initial to 20%	Between 10% points	Base line
Current	0.77τ	2.1τ	2.5τ	3.0τ	3.1τ	3.5τ
Power	0.76τ	1.8τ	2.1τ	2.7τ	2.75τ	3.5τ
Temperature	0.61τ	2.95τ	3.20τ	3.4τ	3.4τ	3.5τ

unit surface area of bore and the Stefan Boltzmann law that power radiated is proportional to the fourth power of the absolute temperature. The curves of Fig. 6.1 show the time development of the normalized current, power input, and temperature, for the optimum damping of $\alpha = 0.8$, from the Markiewicz and Emmett curves and the fourth power law. Note that the power curve is shorter than the current curve. Because of the inherent threshold in the laser action, the duration of the laser output pulse will be even shorter. Extended pulses can be created using sectioned L-C circuits. For reference later, the variously defined pulse durations are listed in Table 6.2 expressed in terms of the electrical time constant τ. Confusion can arise from different pulse definitions in the literature. We will use the full width half maximum (FWHM) value unless otherwise indicated.

6.3.3 Power Supplies and Modulators

The combined cost of the power supply and modulator often dominate, so laser efficiency is important. About 0.5% overall efficiency is typical, and 1.5% is unusually high. The power supply generally converts the alternating line voltage to high voltage, direct current, for charging the storage capacitors between laser pulses. The voltage is generally between 5 kV and 40 kV, and the charging current no more than a few amps per lamp. The trend is toward resonant-switching power supplies, which have become smaller and more reliable in recent years.

Power supply and modulator details can be tricky, especially in the finer points of how the charging and simmer/pre-pulse currents are handled in the modulator. The circuit impedances vary with time during the lamp pulsing and capacitor recharging. Designs tend to be proprietary. Protective arrangements are needed to prevent output under fault conditions, such as open- or short-circuited lamps and connections. Misdirected energy can be disastrous. Careful design and choice of materials are needed to avoid corona, since this produces ozone which attacks insulation and magnifies any incipient problem. Powerful electromagnetic interference is generated, particularly as the discharge starts. Interference is rich in the 10–30 MHz range.

The modulator contains capacitors to store the charge, and circuitry for switching and shaping the discharge into the lamp. The short energetic pulses usually require charging the capacitor bank to a voltage above the breakdown value for the lamp, hence a series switch is needed. A thyratron or triggered spark gap is often used. The modulator may also contain circuitry for coupling the simmer or pre-pulsing current.

The lamp discharge circuit generally consists of the capacitor bank, the lamp, and the switch, in a single closed loop. The circuit is kept as short and simple as possible, since the current may peak at many thousands of amps, and last a few microseconds or less. For safety and engineering simplicity, the common point between the lamp and switch is usually grounded. The capacitor bank is generally charged through an impedance with low frequency-pass characteristics, usually through a rectifier. During the charging, the flashlamp end of the capacitor bank is kept at relatively low voltage, either by suitable impedance to ground, or by virtue of simmer current flowing. When fully charged, the high voltage is at the common point of the capacitor bank and the switch. The switch then closes when the laser pulse is required, rapidly transferring the high voltage (now of opposite sign) to the ungrounded end of the flashlamp, leading to breakdown of the lamp, and discharge of the capacitor bank through the lamp and switch. The lamp cables must have minimum inductance, and may be stripline or shielded coax with length limited to a few feet.

A slightly underdamped discharge circuit is often used. The small current reversal helps extinguish current flow in the switching device. However, too much reversal will reduce the efficiency, and may damage the lamp anode unless designed to also act as a cathode. Xenon Corporation supplies such "double cathode" lamps. Discharge circuits are discussed by *Goncz* [6.78], *Perlman* [6.91], and *Markiewicz* and *Emmett* [6.90]. *Ewanizky* and *Wright* [6.92] describe a Marx-Bank driver circuit for charging capacitors in parallel and discharging them in series.

The high voltage and simmer supplies are isolated during the discharge by impedances with low-frequency pass bands and possibly a rectifier. The arrangement just described lends itself to having a number of lamps controlled by a single switch common to the otherwise individual lamp circuits. Depending on the pulse rate, the simmer may be a few amps continuous, a combination of about 30 mA continuous and a few amps pre-pulse just before the main pulse, or a pre-pulse only. The lamp passes through a negative slope in its impedance characteristic, so the simmer supply should have active current control.

6.3.4 Emission Characteristics

At temperatures below about 10 000 K the lamp plasma is partially transmitting, and there is strong line emission determined by the fill gas [6.93]. When lamps are driven harder, the temperature of the plasma increases, it becomes more opaque and the spectrum more black-body like, determined by the electron temperature in the plasma [6.82, 94]. The opacity and emissivity approach unity over the whole spectrum. For example *Goncz* and *Newell* reported [6.93] that the spectrum of an EG & G FX-47A flashlamp, between 0.3

Table 6.3. Emission from EG & G Lamp FX-47A, pulse length $T = 750\,\mu s$ [6.93]

Energy [J]	Current density [A/cm²]	Equivalent temperature [K]	0.35–0.5 μm	Efficiency % 0.5–0.7 μm	0.7–0.9 μm	0.9–1.1 μm	0.35–1.1 μm total
5000	5300	9400	27	20	11	6	64

and 1.1 μm wavelength, closely approximated a black body at 9400 K when excited with 5300 A/cm². They found the percentages of the excitation energy radiated into various spectral intervals as shown in Table 6.3. Most of the electrical power must have been radiated by the plasma, since some was outside the spectral interval measured (much absorbed by the jacket).

Other investigations which have contributed to the understanding of emission characteristics of lamps are listed by *Everett* [6.74]. Of particular interest, *Gunther* [6.95] developed black-body modeling of flashlamps, and *Gavrilov* et al. [6.96] investigated xenon lamp characteristics at temperatures between 17 000 and 28 000 K. *Giterman* et al. [6.97], while studying gas discharges, achieved 26 000 K quasi-steady state in an air-xenon mixture for pulses lasting up to 100 μs. *Pacheco* et al. [6.73] made quantitative spectral measurements which will be discussed later in this section.

The spectral shape varies during the pulse [6.98, 99]. The spectrum used to characterize the emission is generally obtained by integrating over the entire discharge pulse. The shapes of these spectral curves depend strongly on the ratio E/TA, where E is the discharge energy, T is the pulse duration and A is the bore surface area of the envelope surrounding the arc region [6.82]. In ILC literature, the duration may be taken as the time between initiation and the fall-back to 20% of peak current of the current pulse, or the time between the 10% points. These definitions give $T = 3.0\tau$ or 3.1τ. Elsewhere it is generally taken as the time between the 1/3 peak-current points, giving $T = 2.5\tau$. ILC report that for flashlamps in common use, E/TA is generally about 12.5 kW/cm². We will see later that, for dye lasers, E/TA can exceed 1 MW/cm². Hence, generalizations valid in other fields may not be relevant to FEDLs.

a) Black Body Model

A body in thermal equilibrium can never emit more, in any spectral interval, than a black body at the same temperature. Black body radiation characteristics are well known. The curves of Fig. 6.2 are based on relations from *The Infrared Handbook* [6.100]. A flashlamp will emit as a black body in spectral regions where the plasma is opaque (i.e. emissivity is unity). If the plasma is optically thin, then the atomic spectral lines of the fill gas will shine through. The envelope is generally fused silica which absorbs radiation at wavelengths shorter than about 180 nm, depending on its grade. Spectra of lamps excited at different levels are found in literature from lamp makers.

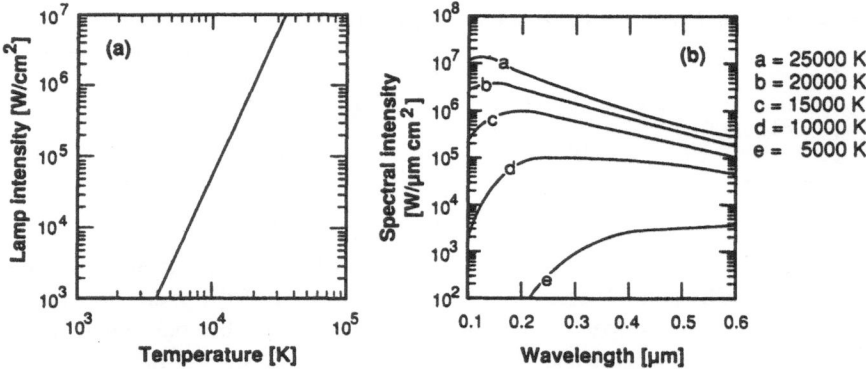

Fig. 6.2. Curves of power radiated by black bodies, (**a**) total intensity vs. temperature, (**b**) spectral intensity vs. wavelength

If a lamp radiates as a black body with radiation balancing input power, then the temperature will be determined by the electrical power per unit area of bore surface, P. For pulse operation we will define a "representative" temperature determined by the "representative" value of P, i.e. P_R. Reference to the power curve of Fig. 6.1 confirms that a rectangular pulse with the same peak power and FWHM would contain the same energy as the actual pulse, to a close approximation. Since the power FWHM is 1.8τ (see Table 6.2), a good choice for P_R is

$$P_R = \frac{E}{1.8\tau A} .$$ (6.2)

Hence, if k_B is the Stefan Boltzmann constant, the representative temperature T_R is given by

$$T_R = \left(\frac{E}{k_B 1.8\tau A} \right)^{1/4} .$$ (6.3)

In as much as a lamp is a black body in quasi-equilibrium, then its "representative" emission characteristics will be governed by the ratio $E/\tau A$; and its "representative" plasma temperature and spectral distributions can, at least approximately, be obtained from the black body curves of Fig. 6.2. We can expect the "representative" characteristics to closely approximate measured integrated ones. Errors in prediction of temperature (and hence spectral distribution) will be smaller than the errors in prediction of power radiated, because of the fourth power relationship.

b) Temperature Related to Current Density

From (6.1), the power input to a lamp is given by

$$I^2 R = \frac{1.13 L_A}{\sqrt{A_C}} I^{3/2} \text{ Watts} . \tag{6.4}$$

Integrating over time to obtain the total energy, and dividing by τA gives the $E/\tau A$ ratio

$$\frac{E}{\tau A} = \frac{1.13 L_A}{\tau A \sqrt{L_A}} \int I^{3/2} \, dt . \tag{6.5}$$

We will now define a "representative" current density by

$$j_R^{3/2} = \frac{1}{2.1\tau} \int \left(\frac{I}{A_C} \right)^{3/2} dt , \tag{6.6}$$

since (2.1τ) is the pulse length FWHM. This gives

$$\frac{E}{\tau A} = \frac{2.37 L_A A_C}{A} j_R^{3/2} . \tag{6.7}$$

With simmer or pre-pulse, the discharge very quickly fills the bore, so the ratio of cross sectional to bore area $A_C/A = r/2L_A$. So

$$\frac{E}{\tau A} = (1.18r) j_R^{3/2} . \tag{6.8}$$

From (6.3), this ratio $E/\tau A$ determines the representative temperature and spectrum, which are thus determined by the lamp radius and the representative current density.

c) Quantitative Comparisons

Pacheco et al. [6.42] excited lamps of 7 mm bore, 457 mm arc length, with 8 μs (FWHM), 440 and 140 J pulses, with simmer. Quantitative spectral measurements matched black bodies at temperatures of 20 000 and 15 000 K. Their pulse length, in Table 6.2, gives $\tau = 3.8$ μs and an electrical power pulse of duration 6.8 μs FWHM. The lamp internal bore surface area was 100 cm². Thus, from (6.2), we infer representative total power P_R of 650 and 210 kW/cm² for the two cases. From the curves of Fig. 6.2a, these estimates correspond to 18 400 K and 13 900 K, i.e. about 8% lower than the temperatures deduced from the spectra. *Pacheco* has suggested [6.101] that the higher temperature may result from the envelope absorbing deep UV light and re-radiating it back into the plasma. Also, some of the radiated light must return to the lamp, either from imperfections in coupling to the dye cell or after being transmitted by the dye. Since the plasma is opaque, this light must be reabsorbed and contribute to higher effective efficiency.

These data suggest that intensely driven lamps may be more efficient than expected for achieving power at the required wavelengths. Simmering, or prepulsing, allows such harder driving and also increases the output in this regime.

Detailed modeling of a high power dye laser using the black body estimate is given in [6.74]. The example used 7 mm bore lamps of length 610 mm excited with 500 J pulses, 2.6 μs FWHM current duration, or 2.2 μs FWHM power duration. This yields a P_R value from (6.2) of 1.7 MW/cm^2, resulting in a temperature estimate of 23 400 K. But detailed power balance indicated that the temperature should have been about 27 000 K. The higher apparent temperature was likewise attributed to reabsorption of light by the lamps, both of the UV and the light returning to the lamp. Black-body modeling seems reasonably accurate providing allowance is made for the light reabsorption.

6.3.5 Unconventional Lamps

While most high power FEDLs publicized recently have used conventional linear flashlamps, other types are worthy of mention. Although not available "off the shelf," they may offer some potential.

a) Ablating Silica Lamps

Gibson [6.28] and *Efthymiopoulos* and *Garside* [6.102] experimented with lamps driven into a higher regime, following the suggestion by *Ferrar* [6.27] that in intensely pumped flashlamps the discharge is carried primarily by ablation products from the wall of the silica tube. They used demountable flashlamps with flowing gas, and drove them beyond the normal explosion limit, using wall thicknesses of a few mm instead of the normal 1 mm. Gibson used lamps with 170 mm discharge length, 5 mm bore, aluminum alloy electrodes, air-fill and current rise time of 1 μs. Each lamp operated successfully for more than 40 000 shots at 250 J input. They found that at these intense loadings the radiation was affected only slightly by the fill-gas composition, as expected if performing as a black body. Their lamps cleaned themselves by blowing the ablated silica and sputtered electrode material out the ends. So the electrode material matters less, and the pressure waves from the excitation can dissipate. The "end of life" was when ablation made the bore too large. Reasonable lives were obtained for excitations above the explosion levels of normal lamps having the same dimensions and pulse durations.

Efthymiopoulos and Garside, using argon fill and 2 mm bore, found that, as energy to the lamp increased, the light intensity in the 0.53 μm wavelength region (approximating the pump band of rhodamine 6G dye) went up with improving efficiency, until it saturated into a black-body behavior in the regime where the wall ablation became significant. This happened at excitation of about 150 J/cm^3 and continued to the highest tested points of 350 J/cm^3. The absorption of energy by the ablating silicon was reported as insignificant. At specific input energies above about 150 J/cm^3 the emitted radiation followed that predicted from a black-body model, but at lower input energies the output was less than predicted. However they did not report using simmer. Its use might have extended the black-body performance to lower specific input energies. We will return to these data below and make estimates of the temperatures achieved. Their lamps sustained more than 5000 pulses even at 350 J/cm^3. Although the makeup of the fill-gas had little effect on the radiative output, they did find an effect on the initial rise time of the pulse, but again this might not have occurred had simmer been used. It is not clear why, with the black body emphasis, the results were reported in terms of energy and discharge volume rather than power and emitting surface area.

Their lamps had estimated E/TA of 6.7 and 15.7 MW/cm^2 for the specific energy inputs of 150 and 350 J/cm^3. Hence the representative surface power densities are estimated at 9.4 and 22 MW/cm^2, leading to temperature estimates lof 36 000 K and 44 000 K. A caveat is introduced however, because they observed a relatively long exponential fall-off in the radiation (but it is not clear whether there might have been a strong infrared component in this decay which would make this less significant). The exponential decay constants were about 23 and 60 μs for the two cases. This might suggest that the radiating times were really about 12 and 30 μs instead of the 0.7 μs inferred from the excitation pulse length, which would reduce the representative power densities to about 530 kW/cm^2 in both cases. The representative temperature would then be 17 500 K. Hence the two extremes of the possible conclusions are: a) extremely high temperatures can be reached by ablating flashlamps or,.b) they top out at a temperature of less than 20 000 K (perhaps the ablation temperature of silica) and then keep radiating the stored energy after the flash is over. If the latter is the case then one might as well use a longer, less intense, excitation.

Okada et al. [6.103] obtained pulses of 23 J, with efficiency of 0.31%, using four wall-ablation lamps in a single head-to-pump rhodamine 6G, with COT as a triplet quencher. They estimated 21 000 K as the equivalent black body temperature of their lamps, and claimed that about 2% overall efficiency should be practical.

b) Vortex Stabilized Lamp

In another lamp capable of handling high power, a fast flowing vortex of gas is maintained. It swirls around in the tube and constrains the discharge to the center. *Mack* [6.104], *Mack* et al. [6.30] and *Morey* and *Glenn* [6.31] have reported lasers pumped with vortex lamps. Mack reported operation at a black body temperature of 70 000 K. As the input energy was varied, the discharge diameter varied such that the 70 000 K temperature was maintained. At this temperature a black body will radiate 0.14 GW/cm^2, but only 4.9% of this will pass through the quartz envelope. Vortex lamps, able to give continuous power of 300 kW, are commercially sold by the Vortek corporation for industrial processing. These have a film of flowing water maintained against the constraining tube by the swirling gas. Such lamps appear promising for very high power applications.

c) Open Discharge Lamp

Holzrichter and *Emmett* [6.105] reported an open discharge lamp with the brightness of a 30 000 K black body, with the light output following the shape of the current pulse. A discharge was generated in helium or argon gas flowing from a 15 mm diameter ceramic tube into the atmosphere. The resulting light was emitted axially from the end of the tube into an $f/8$ cone angle. Presumably the source could have been imaged into perhaps a 2 mm diameter spot using the full solid angle to obtain the 4.6 MW/cm^2 associated with a black body at 30 000 K, but only over a few mm^2. If used with an intermediate spectral-transfer dye, this should allow very good dye efficiency because of the high power density. This type of lamp has interesting possibilities for high power dye lasers if the noisiness were overcome and it were developed to illuminate larger areas.

d) High Pressure Arc Lamp

Thiel et al. [6.32] used modified high pressure arc lamps to pump a dye jet in 1987 and achieved continuous wave operation. While this system was marginal, it does indicate there is still development potential.

e) Pinched Discharge Lamp

Alekseev et al. demonstrated a "pinched discharge" lamp in 1972 [6.106], in a coaxial geometry, with which they obtained 0.2% overall efficiency in an FEDL.

f) Argon Bomb Lamp

Energy can be stored much more densely in chemicals than in capacitors. Various proposals have been made for creating a brightly luminescent shock wave from an explosive charge, for pumping a dye laser. Approaches for such "argon bombs" have been suggested by *Held* [6.107] and by *Schäfer* [6.108]. Such devices have obvious military interest.

g) Potential Improvements in Efficiency

Baranov et al. [6.109] demonstrated that the efficiency of a lamp could be increased by having an interference dielectric coating on the envelope, to return light having wavelengths not absorbed by the medium, so it might be re-absorbed by the plasma. Their application was in the pumping of Nd:YAG lasers, but the principle should also apply to the shorter wavelengths more often used for dye lasers.

Since dye lasers have a substantial threshold in their operation, the output energy increases at a higher rate than the pumping energy. This encourages driving the lamps hard. At the resulting high temperatures, much of the energy may be at wavelengths too short to be absorbed by the lasing dye. This gives a strong incentive to use spectral conversion by intermediary dyes. This can substantially increase the effective 2.9% lamp efficiency of Table 6.1. Such an agent may be in the envelope of the lamp, or at some other point between the lamp and lasing dye. This approach will be discussed further in the dye section.

h) How Critical is the Fill?

If a lamp is in the black-body regime, with input power balanced by radiation, the details of the fill should be unimportant, provided the discharge can be initiated and sustained. Hence, in such a lamp the choice of fill will probably have more to do with impedance matching and breakdown characteristics than with spectral details and efficiency.

6.3.6 Lamp Degradation

When lamps age, the following happens [6.82]. Chemical and metallic deposits from the electrodes form on the inside of the envelope, blocking emission. The quartz envelope cycles through high temperatures, resulting in vaporization, melting, ablation and crystallization. This can lead to blocking of light, crazing or cracking. Contamination of gas fill can make triggering or simmering difficult and cause failure of electrode seals. Crystallization (or devitrification) begins at the inside surface and results in cristobalite, an opaque form of SiO_2. The process can accelerate since absorption of the light further heats the surface [6.110]. Impurities in the quartz envelope, and solarization caused by UV action on color centers, can also absorb light. These effects can be annealed out at temperatures over 500°C. In some cases lamps self-anneal during lamp operation. The prevalence of solarization depends on the quality of the quartz and on the operating conditions.

a) Expected Lamp Life

Lamp life is a strong function of energy applied and length of the pulse. *Goncz* [6.78] reported the following empirical expression (translated to metric units) for energy E_x that will just explode the lamp. It was derived from testing lamps with pulse time constants varying from 5 μs–3 ms, bore diameters varying from 4 mm–28 mm, and various lengths.

$$\frac{E_x}{L} = 112D\sqrt{T} \text{ J/mm} , \tag{6.9}$$

where D is the bore diameter in mm, L is the lamp length and T is the duration of the discharge pulse between 1/3 maximum current points (in seconds). This rearranges to give

$$\frac{E_x}{A} = 3570\sqrt{T} \text{ J/cm}^2 \, , \qquad (6.10)$$

where A is the bore surface area (cm^2). Thus, the sustainable energy density on the bore of the lamp is proportional to the square root of the pulse duration.

Goncz also reported that life testing with varied lamps gave the results shown in Table 6.4 [6.78], in which the flash energy is normalized to the explosion energy. He noted that lamps in enclosed cavities, and helical lamps, should be derated since some of the energy reflects back to the lamp.

The above data have been amplified in EG & G data sheets [6.111]. The EG & G data are in general agreement with (6.10) and Table 6.4, and provide a guide to ultimate energies and life expectations for a wide range of lamp sizes and pulse lengths with wall thickness of 1 mm. The minimum and maximum life limits are given approximately by

$$\left(\frac{E_0}{E_x}\right)^{-5.5} < N_L < \left(\frac{E_0}{E_x}\right)^{-11} \, , \qquad (6.11)$$

where N_L is the number of pulses before the output drops by 50%. The data was generated for pulse lengths between 10 μs and 10 ms, so does not quite cover our regime. However, the extrapolation is small, with no suggestion of departure down to 10 μs, and it is the best general data we have. Derating is required if light is reflected back to the lamp.

ILC report [6.82] that N_L can be obtained from the relation

$$N_L = \left(\frac{E_0}{E_x}\right)^{-8.5} \, , \qquad (6.12)$$

where N_L is the number of pulses before the output drops by 30%. This is consistent within the uncertainty arising from the different definition of N_L and of pulse duration. ILC does restrict the relation to lamps of less than 15 mm bore (they report an inverse 14th power for larger diameters). Further such data are discussed in [6.74].

Table 6.4. Expected flashlamp life [6.78]

Flash energy [percent of E_x]	Flashlamp life [number of flashes]
100%	0–10
70%	10–100
50%	100–1000
40%	1000–10 000
30%	10 000–100 000

b) Recent Experimental Data

More recent experiments have shown lamps lasting longer than would be expected from the above data. Figure 6.3 shows the degradation curve for a simmered linear xenon lamp running at 59% of the explosion level, as calculated by (6.10) [6.41], using 410 J pulses lasting 3.2 µs. The expected life N_L to 50% output, from (6.11), would lie between 30 and 300 pulses. But it showed very little deterioration after 300 000 pulses. The output was monitored by illuminating a dye cell containing rhodamine 6G and monitoring the fluorescent output.

The tendency for lamp output to recover between spells of use can be observed. This overnight recovery effect is well known to users of dye lasers, but is often attributed to dye recovery. Why it occurs is not clear. While the recovery effect may be only about 10% in useful light output from the lamp, it can translate into a much larger change in laser output, particularly if operating near threshold.

In an operating laser system discussed earlier [6.74], simmered linear xenon lamps are run at 59% of the explosion level of (6.10), with 500 J pulses of 2.6 µs. There has been little loss of output after thousands of pulses, even though the lamps are enclosed in a laser head.

c) Discussion of Lamp Life

The reports just referred to, and also the discussion by ILC [6.82], and by *Furomoto* [6.6], indicate that flashlamps in this pulse regime may be driven much harder than was previously thought, provided they are simmered or pre-pulsed. Such good life can only be expected when the lamps are produced under excellent quality control to the best available design, and then are used under

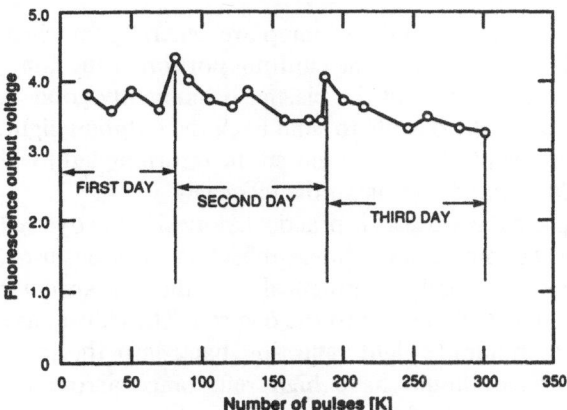

Fig. 6.3. Lamp degradation when repetitively pulsed at 59% of "explosion" energy [6.41] with 410 J, 3.2 µs pulses, monitored by fluorescence from a cell of rhodamine 6G

ideal conditions. Premature failure can be caused by contamination of fill, electrodes or envelope; by incorrect handling, mounting or cooling; or by non-optimal excitation or simmering. If there is anything wrong, the laser tends to die rather suddenly. Life-testing data are expensive to generate and published reports are rare. Laser output usually degrades faster than that from the lamps, particularly if working close to lasing threshold.

Our later modeling will show that, even with a good dye such as rhodamine 6G, a practical laser needs at least $10\,kW/cm^2$ into the pump band to be efficient, and more is desirable. Since it is difficult to achieve even 10% of the total lamp energy coupled into the dye pump band, this implies that the lamp loading should certainly exceed $100\,kW/cm^2$, and preferably exceed $1\,MW/cm^2$. For a given power output, the size of the laser is also governed by the lamp power density, so there is always incentive to drive the lamps as hard as possible compatible with desired life. If the technology develops to operate at even higher temperatures, then the spectral shift to shorter wavelengths will provide a yet stronger push to spectral conversion by an intermediary dye, a subject to be discussed later.

d) Lamp Cooling

Water cooling is generally used. Practical considerations generally result in the discharge voltage being across the cooling water, as well as across the lamp, so the water resistivity must be controlled. At least $0.5\,M\Omega$ cm is considered desirable. Values over $1\,M\Omega$ cm can cause etching of metallic components in the system. Air cooling can be practical for low pulse rates.

6.4 Coupling the Flashlamp to the Dye

For efficient coupling, lengths of dye cell and flashlamp are generally matched. With coaxial lamps, the dye is surrounded by the emitting portion of the lamp, and the coupling from the interior surface of the plasma is inherently good. A reflector is generally placed around the outside to send back the outgoing light. If the lamp is driven hard, then the plasma is opaque, so the returning light will be reabsorbed, heating the plasma and so raising the efficiency.

Linear lamps are usually placed as close as is practical alongside the dye cell within a surrounding reflector. In some cases a diffuse reflecting surface is used, surrounding lamp and dye cell as closely as practical. In others, a specular elliptical reflector is used, imaging each lamp into the dye cell. The ellipse may have detailed modifications to minimize light reflection back into the lamp [6.82]. In either case the surface should have high reflectance across the spectrum that includes the pump band of the dye (and possibly any spectral transfer agent). It must withstand intense light without deteriorating. Diffuse reflectors have included barium sulphate and Teflon. Specular reflectors have included aluminum and silver, usually overcoated for enhancement and protection, or placed on the outside surface of glass or fused silica. A spherical reflector housing has also been used with a vortex lamp, with the lamp and dye cell placed on a single axis, imaged into each other [6.104]. *Basov* and *ILC* describe a number of coupling arrangements in more detail [6.5, 82].

With an elliptical specular reflector, the geometrical efficiency can theoretically be 100% if the dye cell diameter is no less than the lamp diameter, and the reflector size is relatively large. If the dye cell is smaller, the coupling efficiency falls linearly with its diameter. Efficiency drops further with eccentricity of the ellipse [6.82, 112]. Coupling efficiency is further lowered if the specular reflector does not maintain high reflectance over the wide ranges of spectrum and angle of incidence. The highest reflectances are obtained with multi-layer dielectric coatings, but these are affected most by changing spectrum and angle of incidence.

Close-coupling efficiency can be no higher than the ratio $A_D^2/(A_D^2 + A_L^2)$ where A_D and A_L are the total surface areas of the dye cell and lamps, respectively. The equality holds in the case of perfect diffuse reflectors with complete absorption of light falling on lamps and dye cell. It comes from the "brightness theorem," which states that no passive optical system can increase brightness, which is defined later in (6.43). In principal, a small gain could be made by judicious choices of refractive index, but this opportunity is generally limited. Coupling efficiency can thus increase towards 100% with a relatively large dye cell. But the rate of improvement lessens when the dye cell surface area exceeds that of the lamps; whereas, the power intensity continues to drop. In practical close-coupled lasers, the dye cell surface area is generally quite closely matched to the total emitting area of the lamps because high power density is important, as discussed above. This holds for slab dye lasers and for those with cylindrical cells. The optics of non-imaging systems are discussed by *Weiner* [6.113] and *Welford* and *Winston* [6.114].

Many lasers use close coupling with diffuse reflectors, although efficiency with specular imaged coupling is theoretically higher. This is because materials are available with high diffuse reflectance over broad spectral ranges and angles of incidence. Barium sulphate powder in a liquid slurry, such as that obtainable from Eastman Kodak Company, is often used. Such coatings can readily be replaced if damaged.

It has been supposed that the large dye cells associated with dye lasers are needed to contain enough dye to emit the energy involved, but this is erroneous. The author has seen 180 mJ emitted in a 2.5 μs pulse from 0.25 cm^3 of sulforhodamine 640 dye when pumped with an FEDL laser using coumarin 504. The relatively large dye cells are needed for efficient coupling of the lamps into the dye. The need to match dye cell dimensions to those of the lamps can lead to pumping undesirably large volumes of dye. An advantage of the one-dimensional waveguide dye cell is that it increases the ratio of surface area to volume. Such a cell may be less than one mm in the waveguide dimension, but a few cms in the other transverse dimension [6.115–118]. With this approach less dye has to be moved to obtain clearing of the dye cell between successive lasing pulses. For many applications, the resulting aspect ratio of the output beam is awkward. However, anamorphic optics (such as astigmatic eye-glass lenses) can be used to change the aspect ratio.

6.4.1 UV Filtering

Pyrex is often chosen for coolant tubes surrounding flashlamps because it absorbs some of the near-UV spectrum which is generally not very beneficial in pumping, but damages the dye and shortens its life. Sometimes a filtering agent is included in the cooling water. When filtering is not desired then quartz is often used for coolant tubes.

6.5 The Dye

For the photochemistry of dyes, the reader is referred to other references, such as a recent discussion by *Jones* in [6.8]. Here we will concentrate on the various transitions that drive the laser, or detract from its operation.

A first-order rate analysis by *Siegman* [Ref. 6.2, p. 39ff] shows that the saturated exponential gain g_L is related to the small signal value g_{L0} by relation

$$g_L = \frac{g_{L0}}{1 + I_L/I_{sat}} , \tag{6.13}$$

where I_L is the lasing photon intensity, and I_{sat} is its saturation value at which g_L has dropped to $g_{L0}/2$. Also, the ratio R of induced to spontaneous transitions is given by

$$R = I_L \tau_{eff} \sigma_{LE} = I_L/I_{sat} , \tag{6.14}$$

where τ_{eff} is the natural lifetime of the excited state, and σ_{LE} is the cross section for stimulated emission. Note also

$$I_{sat} \approx \frac{1}{\tau_{eff} \sigma_{LE}} \text{ photons cm}^{-2}\text{s}^{-1} . \tag{6.15}$$

It follows that the quantum efficiency η_L of converting absorbed pump photons into lasing photons has an upper bound given by

$$\eta_L \leqslant \frac{1}{1 + I_{sat}/I_L} . \tag{6.16}$$

Efficiency requires $I_L > I_{sat}$, to ensure excited molecules are stimulated down before they decay spontaneously. High gain, on the other hand, requires much upper state population, leading to high spontaneous emission. This is particularly significant in dyes because I_{sat} is generally high, about 1.5 MW/cm^2 for rhodamine 6G. The following analysis will develop a fuller picture of what efficient lasing requires.

We will find conditions for efficiently converting pump photons into lasing ones, introducing nine dimensionless parameters p_{ij}. Useful insight will come from an approximation, valid when all p_{ij} values are small, giving the lasing quantum efficiency $\eta_L \approx 1 - \Sigma p_{ij}$. Each p_{ij} is associated with a loss mechanism, and always contains a ratio between two dye parameters, usually modified by the lasing intensity, the pumping intensity, or a ratio of them. Thus η_L can be optimized by choice of the intensities, taking the dye parameters into account. But first we will summarize the spectroscopy.

6.5.1 Dye Spectroscopy

The discussion will assume familiarity with basic laser physics, as in Chap. 1 of *Siegman's* book [6.2] on lasers, his discussion of cross sections and transition probabilities on page 293, and gain on page 287. The reader is also referred to Chaps. 1 and 5 of *Schäfer's* book [6.1] for detailed understanding of the physics, dye chemistry and measurement techniques. The critical concepts will be summarized here.

Population is excited from relatively low vibrational levels of the ground electronic state, into the vibrational manifold of a higher electronic state. These manifolds typically thermalize within picoseconds to a thermal distribution in the lower vibrational levels of the excited state [6.1, 2]. The population inversion needed for optical gain occurs between these levels and the less populated upper vibrational levels of the ground electronic state. The destination levels deplete so rapidly that bottle-necking is usually insignificant. Rate equation analysis allows good understanding of the processes.

Since the upward transitions go through a larger energy difference than the downward ones, there is an offset, typically about 30 nm, between the spectral distributions of absorption and spontaneous emission. The broadenings are generally asymmetric, with the spontaneous emission approximately a mirror image of the absorption. The gain will tend to be higher for transitions into unoccupied levels, giving a further offset dependent on the levels available. Typically this lasing shift is about 50 nm from the absorption peak, unless purposely tuned by introducing a frequency-dependent element. The spectra are normally "homogeneous" because of the rapid thermalization, i.e. the population rapidly redistributes in any vibrational complex to fill "holes" in population left by an outgoing transition. Thus "hole-burning" in the linewidth, from intense narrowband radiation, will not be observed unless the resultant transition rate exceeds the thermalization rate (a rare occurrence).

6.5.2 Significant Transitions

The significant transitions are shown in the eigenstate diagram of Fig. 6.4, modeled after *Schäfer* [6.1] and *Snavely* [6.119]. The electronic states of interest are the ground, the first and second singlet, and the first and second triplet, with population densities N_0, N_1, N_2, N_T and N_{T2}. Each electronic state has its vibronic-rotational complex. The pumping and lasing intensities I_P and I_L are involved in radiative transitions between the ground and singlet states, and can also be absorbed while causing transitions to higher states. The transitions are divided into three groups; arising from the pump, from the lasing flux, and from spontaneous transitions. The spontaneous rates γ_{10} and γ_{1T} from the singlet are generally of the order of 10^8/s, but the triplet emptying rate γ_{T0} tends to be much slower, about 5×10^5/s for rhodamine 6G [6.1]. This allows population to build up in the triplet state. The resulting absorption of light by the triplet state generally becomes a serious loss for laser pulses lasting more than a few hundred nanoseconds, unless an effective triplet quencher is included. Fortunately oxygen, which is present anyway when the dye solution is equilibrated with air, is a triplet quencher for at least some dyes. But it may also increase the triplet crossing rate γ_{1T} so is not always beneficial. The important subject of triplet quenching is discussed by *Schäfer* [6.1]. Excitations into higher singlet and triplet states generally return non-radiatively to their respective singly excited levels with high transition rates, of the order of 10^{12}/s. We will assume the relaxation is immediate.

6.5.3 Cross Sections and Wavelength Dependence

If $\sigma_{ij}(\lambda)$ were the radiative cross section for transition at wavelength λ between individual electronic-vibrational states i and j, then we would expect $\sigma_{ij}(\lambda) = \sigma_{ji}(\lambda)$. But, to use this reciprocal relationship, we would need to keep track of the populations in all the vibrational levels of all the significant electronic states, which would be too complex. Our analysis will keep track only of the total

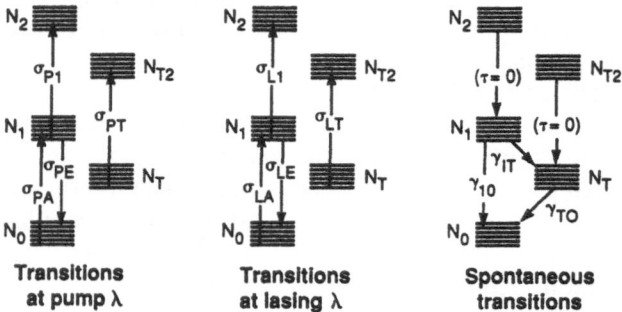

Fig. 6.4. Schematic energy level diagram for a dye molecule, showing transitions of interest [6.1]

population in each of the electronic states, N_i being the population density in electronic state i, and $\sigma_{ij}(\lambda)$ being the radiative cross section between electronic states E_i and E_j. We will thus lose the direct reciprocal relationship. But, generally, the values $\int \sigma_{ij}(\lambda)d\lambda$ and $\int \sigma_{ji}(\lambda)d\lambda$ over each line will be approximately equal, albeit displaced by maybe 30 nm. As temperature changes, the thermalized distribution in the vibronic complexes will vary somewhat and affect the measured values of $\sigma_{ij}(\lambda)$. But this is normally a small effect.

We must allow for the wavelength dependence of each of the cross sections. For our initial analysis, we will assume that the pumping and lasing fluxes are each at a discrete wavelength. Each cross section will then take the value for the relevant wavelength, and hence have different values for the pumping and lasing. The spontaneous transitions radiate into a wavelength distribution, but we are concerned only with the total probability.

6.5.4 Labeling

We will follow *Siegman's* use of symbols and units [6.2] with a few exceptions. It is more practical to measure optical power than amplitude, so our exponential gain coefficient g will be for intensity, and will be double Siegman's coefficient α for amplitude. The conventional use of σ_{ij} for transition cross sections between states i and j would be clumsy here. We will use a scheme in which the first subscript is P if relating to the pumping, and L if to the lasing. The second subscript will be A or E to denote an absorbing or emitting transition between the ground and singlet states, and 1 or T if an absorption from the singlet or triplet to the next higher state. We will keep track only of the total population density in each of the electronic states.

The effective lifetime τ_{eff} of the first singlet is related to the spontaneous transition probabilities by $\tau_{\text{eff}} = 1/\Sigma\gamma_{1j}$, where $\Sigma\gamma_{1j}$ is the sum of the spontaneous transition probabilities from the singlet. In practice the sum is generally dominated by γ_{10}, the rate to ground (mostly radiative), so

$$\tau_{\text{eff}} \approx \frac{1}{\gamma_{10}} \ . \tag{6.17}$$

6.5.5 Fluorescent and Lasing Quantum Efficiencies

The fluorescent quantum efficiency η_F is the ratio of number of photons emitted by singlet-ground transitions to pump photons absorbed in the absence of lasing. If γ_{10} is purely radiative, then with the transitions shown in Fig. 6.4, $\eta_F = \gamma_{10}/(\gamma_{10} + \gamma_{1T})$. In the case of rhodamine 6G, γ_{10} and γ_{1T} are about 2.9×10^8/s and 2.0×10^7/s. Hence, $\eta_F \approx 0.93$, which is close to the 0.92 cited by *Snavely* [6.120] as having been measured by Grum at Kodak Research Laboratories. In some dyes γ_{10} has a strong non-radiative component, leading to a lower fluorescent quantum efficiency.

When a laser is working efficiently, the lasing transitions from the singlet to the ground state exceed the spontaneous ones; thus the lasing quantum efficiency can exceed the fluorescent quantum efficiency.

6.5.6 Dimers

Many dyes form dimers, especially in aqueous solutions. The dimers tend not to lase well and may absorb at the monomer's lasing wavelength. Addition of a detergent, such as ammonyx LO, often reduces this aggregation and may assist in quenching triplets [6.1].

6.6 Rate Equation Analysis

The rate equation approximation applies whenever the rate of change of population is slow compared with the reciprocal of the natural linewidth. The condition will be met in most cases of interest for high-power FEDLs, since dye natural line widths tend to be about 2×10^{13} Hz, and typical pumping cycles are longer than 100 ns. *Siegman* gives a fuller discussion of the limits of rate analysis [6.2].

The transitions essential to laser operation involve the pumping cross section σ_{PA} and the lasing cross section σ_{LE}. The spontaneous transition rate γ_{10} to the ground state is a major interfering factor, and is inherently associated with the other two [6.121]. We will call these transitions the "basic" ones. All the other transitions illustrated in Fig. 6.4 are detrimental to the process. We want to minimize their effect by choice of dye and operating conditions.

We will consider an element of dye exposed to pumping and lasing fluxes. A simple picture will first be obtained by including only the "basic" transitions, using discrete frequencies for pumping and lasing. We will then develop the analysis to include all transitions shown in Fig. 6.4, and then further develop it for handling pump power with a distributed spectrum, as occurs in flashlamp-pumped lasers. We will develop relations for gain, saturation, and lasing quantum efficiency. The same framework will allow adding further loss mechanisms if thought significant. We will use conventional rate analysis, following the approach of *Ganiel* et al. [6.122] and *Hargrove* and *Kan* [6.123], assuming quasi-steady state, valid because the transition rates will be much faster than the rate at which the pumping flux will change. The possible exception is the triplet-emptying rate γ_{TO}. But there will normally be a quenching agent (such as oxygen) present in a practical laser, making γ_{TO} sufficiently high to validate the assumption. Homogeneous broadening will be assumed. To keep equations uncluttered with $h\nu$ we will use photon intensities (i.e. photons $cm^{-2}s^{-1}$) or power densities, as convenient.

6.6.1 Gain and Saturation – Basic Transitions

In quasi steady-state, the populations will remain almost constant on the time scale of the various transition rates, so conventional rate analysis yields, with only the basic transitions

$$\frac{N_1}{N_0} = \frac{I_P \sigma_{PA}}{I_L \sigma_{LE} + \gamma_{10}} \, , \tag{6.18}$$

where I_P and I_L are the pumping and lasing intensities (photons $cm^{-2} s^{-1}$). The lasing exponential gain coefficient is

$$g_L = N_1 \sigma_{LE} \, , \tag{6.19}$$

so

$$g_L = \frac{N_0 \sigma_{LE} I_P \sigma_{PA}}{I_L \sigma_{LE} + \gamma_{10}} \tag{6.20}$$

and

$$g_{L0} = \frac{N_0 \sigma_{LE} I_P \sigma_{PA}}{\gamma_{10}} \tag{6.21}$$

is the small signal gain (putting $I_L = 0$). Therefore, gain saturation is described by

$$\frac{g_L}{g_{L0}} = \frac{1}{1 + 1/p_{S1}} \, , \tag{6.22}$$

where

$$p_{S1} = \frac{\gamma_{10}}{I_L \sigma_{LE}} \, . \tag{6.23}$$

The dimensionless parameter p_{S1} is the ratio between spontaneous and stimulated transitions to the ground state, and has value unity when the exponential gain has dropped to half the small signal value. Since I_{sat} is the lasing intensity that causes the gain to drop to half of its small signal value, then

$$I_{sat} = \frac{\gamma_{10}}{\sigma_{LE}} \text{ photons } cm^{-2} s^{-1} \, . \tag{6.24}$$

Note also

$$p_{S1} = \frac{I_{sat}}{I_L} \, . \tag{6.25}$$

With this substitution, (6.22) is the same as (6.13).

The exponential gain seen by the pumping photons, ignoring bulk absorption because of the typically short pumping path, will be

$$g_P = - N_0 \sigma_{PA} .\tag{6.26}$$

We expect g_P to be negative since pumping photons are absorbed.

6.6.2 Lasing Quantum Efficiency – Basic Transitions

Suppose lasing intensity I_L is flowing in direction z, and pumping intensity I_P is flowing in direction x. Then the photons dN_L added to the lasing flux, and the photons dN_P removed from the pumping flux, in the element of dye, are, respectively

$$dN_L = dI_L(dx\,dy) = g_L I_L\,dz(dx\,dy)\tag{6.27}$$

and

$$dN_P = - dI_P(dy\,dz) = - g_P I_P\,dx(dy\,dz) .\tag{6.28}$$

Hence, the lasing quantum efficiency is

$$\eta_L = \frac{dN_L}{dN_P} = - \frac{g_L I_L}{g_P I_P} ,\tag{6.29}$$

and from (6.20) and (6.26)

$$\eta_L = \frac{I_L \sigma_{LE}}{I_L \sigma_{LE} + \gamma_{10}} .\tag{6.30}$$

Equation (6.23) then gives

$$\eta_L = \frac{1}{1 + p_{S1}} .\tag{6.31}$$

Thus, noting (6.25), the above is consistent with the earlier (6.16). If the flow directions are chosen differently, the same relations will always be obtained.

6.6.3 Gain and Saturation, Including Other Losses

We will now expand these relations for gain and efficiency, adding the other transitions in Fig. 6.4. The dimensionless p_{ij} parameters of Table 6.5 will keep appearing. The rate equations now give, for quasi steady-state of the first singlet population,

$$\frac{N_1}{N_0} = \frac{I_P \sigma_{PA} + I_L \sigma_{LE}}{I_L \sigma_{LE} + I_P \sigma_{PE} + \gamma_{10} + \gamma_{1T}} ,\tag{6.32}$$

Table 6.5. List of dimensionless parameters

Symbol	Description	Parameter	Power dependence
p_{S1}	Spontaneous singlet	$\gamma_{10}/I_L\sigma_{LE}$	$1/I_L$
p_{ST}	Spontaneous triplet	$\gamma_{1T}/I_L\sigma_{LE}$	$1/I_L$
p_{L1}	Lasing absorption singlet	σ_{L1}/σ_{LE}	None
p_{LT}	Lasing absorption triplet	$\sigma_{LT}\gamma_{1T}/\sigma_{LE}\gamma_{T0}$	None
p_{L0}	Lasing absorption ground	$I_L\sigma_{LA}/I_P\sigma_{PA}$	I_L/I_P
p_{LB}	Lasing absorption bulk	$kI_L/N_0I_P\sigma_{PA}$	I_L/I_P
p_{P1}	Pump absorption singlet	$I_P\sigma_{P1}/I_L\sigma_{LE}$	I_P/I_L
p_{PT}	Pump absorption triplet	$I_P\sigma_{PT}\gamma_{1T}/I_L\sigma_{LE}\gamma_{T0}$	I_P/I_L
p_{PX}	Pump stimulated singlet	$I_P\sigma_{PE}/I_L\sigma_{LE}$	I_P/I_L

i.e.

$$\frac{N_1}{N_0} = \frac{I_P\sigma_{PA}}{I_L\sigma_{LE}}\frac{1+p_{L0}}{(1+p_{S1}+p_{ST}+p_{PX})} \tag{6.33}$$

where the terms p_{ij} are defined in Table 6.5. In quasi-steady-state, for the triplet population N_T (no longer zero) we have

$$\frac{N_T}{N_0} = \frac{\gamma_{1T}}{\gamma_{T0}}\frac{N_1}{N_0} . \tag{6.34}$$

The exponential lasing gain becomes

$$g_L = N_1(\sigma_{LE} - \sigma_{L1}) - N_T\sigma_{LT} - N_0\sigma_{LA} - k . \tag{6.35}$$

Eliminating N_1 and N_T with (6.33, 34), and making p_{ij} substitutions, we obtain

$$g_L = \frac{N_0I_P\sigma_{PA}}{I_L}\left[\frac{(1+p_{L0})(1-p_{L1}-p_{LT})}{(1+p_{S1}+p_{ST}+p_{PX})} - p_{L0} - p_{LB}\right] . \tag{6.36}$$

This reduces to the earlier "basic" (6.20) when the p_{ij} terms other than p_{S1} go to zero.

6.6.4 Pumping, Including Other Losses

The gain seen by the pumping photons, again ignoring bulk absorption, and expecting it to be negative is

$$g_P = N_1(\sigma_{PE} - \sigma_{P1}) - N_T\sigma_{PT} - N_0\sigma_{PA} . \tag{6.37}$$

Again using (6.33, 34) to eliminate N_1 and N_T, and making p_{ij} substitutions we obtain

$$g_P = -N_0 \sigma_{PA} \left[1 - \frac{(1 + p_{LO})(p_{PX} - p_{P1} - p_{PT})}{(1 + p_{S1} + p_{ST} + p_{PX})} \right]. \tag{6.38}$$

This reduces to the "basic" (6.26) if the p_{ij} terms other than p_{S1} go to zero.

6.6.5 Lasing Quantum Efficiency with Other Losses

The ratio of lasing photons gained to pumping photons lost is still given by (6.29). Hence, from (6.36) and (6.38), making the p_{ij} substitutions, we obtain the rate equation result with no additional approximations

$$\eta_L = \frac{(1 + p_{LO})(1 - p_{L1} - p_{LT}) - (p_{LO} + p_{LB})(1 + p_{S1} + p_{ST} + p_{PX})}{(1 + p_{S1} + p_{ST} + p_{PX}) - (1 + p_{LO})(p_{PX} - p_{P1} - p_{PT})}. \tag{6.39}$$

If we now assume that all $p_{ij} \ll 1$, we obtain the first order solution

$$\eta_L \approx 1 - p_{S1} - p_{ST} - p_{P1} - p_{PT} - p_{L1} - p_{LT} - p_{LB}. \tag{6.40}$$

This tells us whether any particular loss is significant. Later we will find that, even with a good dye, it is hard to ensure that p_{S1} and p_{LT} satisfy the smallness criteria. Also, p_{LB} increases as the dye degrades. If we assume that all the other p_{ij} are small, then we obtain the intermediate approximation, which is easier to use than the full (6.39)

$$\eta_L \approx \frac{1 - p_{L1} - p_{LT} - p_{LB} - p_{S1}(p_{LO} + p_{LB}) - p_{LT}p_{LO} - p_{LB}p_{PX}}{1 + p_{S1} + p_{ST} + p_{P1} + p_{PT}}. \tag{6.41}$$

Note that each of the loss parameters is made up of *ratios* between pairs of cross sections and transition rates, usually modified by the ratio I_L/I_P or by I_L alone. *Never does a single dye property appear on its own.* Also, different dyes may show optimal performances with different values of I_L and I_P. So, we should beware of "comparisons" between dyes which neglect re-optimization of both I_L and I_P.

6.6.6 Effect of Dye Concentration

Note from (6.41) that the lasing quantum efficiency is largely independent of the dye concentration. The only loss parameter in Table 6.5 containing a population density is the bulk loss p_{LB}. Even that dependency may go if k is proportional to N_0, as it may if the dye molecules are the source of the bulk loss. Hence, as far as the lasing quantum efficiency is concerned, the dye concentration may generally be kept as an open parameter for optimizing the pump absorption profile. If too concentrated, the innermost dye molecules will not be reached by the pumping photons, causing inefficiency. The uneven heat deposition will also upset beam quality and tuning capability. If too dilute, the pump energy will pass right through. The optimum concentration will be inversely proportional to the pumping depth. In summary, the dye concentration should be optimized for absorbing as many of the pump photons as possible without sacrificing beam quality from uneven heating, or driving the pump intensity too low at the far side. The latter is easier to satisfy if pumping is from both sides, or the light is reflected back with a mirror. Having optimized the concentration in this manner, the pump and lasing intensities can then be optimized as guided by (6.40) or (6.41).

6.6.7 Generalization to Spectrally Broad Pump

So far, discrete optical frequencies have been assumed for all fluxes. This is legitimate for lasing intensity I_L, but not for I_P when flashlamp pumping. We can generalize for lamp pumping by substituting $\int I_P(\lambda)\sigma_{ij}(\lambda)d\lambda$ in Table 6.5 wherever the product $I_P\sigma_{ij}$ appears. The spectral dependence of any cross section $\sigma_{ij}(\lambda)$ will generally vary more rapidly than that of the lamp's spectral photon intensity $I_P(\lambda)$, so an approximation for the integral may be made. Equate the integral to $I'_{P\lambda}(\lambda)\sigma'_{ij}(\lambda)\Delta\lambda$ with the components as follows. $I'_{P\lambda}(\lambda)$ is the local value of the spectral pumping intensity in the regime of the spectral feature, $\sigma'_{ij}(\lambda)$ is the peak of the spectral feature, and $\Delta\lambda$ is its bandwidth (FWHM).

6.6.8 Discussion

As the dye solution ages in the intense illumination, it tends to form products that absorb at the lasing wavelength. This loss is described by P_{LB}. Its impact will depend upon the ratio I_L/I_P (see Table 6.5), and will reduce the output proportionately faster if the efficiency is already low due to other factors. This can be seen by differentiating (6.41) to obtain

$$\frac{d\eta_L}{dp_{LB}} = -\frac{1 + p_{S1} + p_{PX}}{1 + p_{S1} + p_{ST} + p_{PI} + p_{PT}} \ . \tag{6.42}$$

Thus the degradation rate will depend upon operating conditions. This may contribute to differing conclusions on dye life obtained by different experimenters.

An ideal dye would have a working range of I_L and I_P for which all $p_{ij} \ll 1$. The designer would then expose the dye only to those intensities to obtain high efficiency. The penalty of small departures would be given by (6.40). In practice, all known dyes depart from the ideal, and lasing quantum efficiency is generally less than 50% for practical ranges of I_L and I_P when flashlamp pumping. While (6.40) becomes less accurate, it will still give a qualitative estimate of which losses are important. If all the significant parameters are known, then (6.41) should allow a quite accurate forecast of lasing quantum efficiency. For the most precise optimization the accuracy of (6.39) will be needed. The relations lend themselves to manipulation in a personal computer.

When pumping with a laser, the high brightness allows almost unlimited choice of I_P, and I_L can be controlled by choice of feedback in an oscillator, or input signal in an amplifier. The limits will come from damage or non-linear effects. Flashlamp pumping generally results in I_P less than optimum. It is still possible to obtain high values of I_L, but the resulting high ratio I_L/I_P may increase the parameters P_{L0} and P_{LB} enough to reduce efficiency.

The results appear to ignore the gain. This is because, in an efficient (i.e. saturated) laser, the energy added to the input, i.e. the product of the absorbed pump energy and the lasing efficiency, is the more meaningful parameter. For an oscillator, the input is zero, and the lasing intensity is controlled by the feedback.

Until recently, lack of information on dye constants has limited the value of modeling. However, many of the values can now be obtained using the approach of Chap. 3. The correlation between this modeling and that of Chap. 3 is discussed in Appendix A. Thus dye comparison and laser optimization can now be less empirical. A number of additional resources for dye information are listed by *Everett* [6.124].

The highest overall efficiency that the author has found cited for an FEDL is 1.8% [6.36]. Much higher dye lasing efficiencies are available with laser pumping. For instance, more than 60% has been reported when pumping rhodamine 640 with a copper vapor laser [6.125]. More than 40% has been achieved when pumping dye with an excimer laser [6.126], and the author has seen over 40% when pumping rhodamine 6G with an FEDL using coumarin 504. *Bos* [6.127] has reported 55% efficiency when pumping rhodamine 6G with a frequency-doubled Nd: YAG laser. These efficiencies are from the laser pump to the output and ignore losses in the pump laser. The overall efficiency will generally be less than with flashlamp pumping.

6.6.9 Example: Analysis for Rhodamine 6G

We will use rhodamine 6G as an example, because it is the best documented laser dye. The values of the most important parameters are plotted against wavelength in [Ref. 6.1, p. 93]. The parameters are listed in Table 6.6 [6.128, 129] for 530 nm wavelength, corresponding to peak $\sigma_{PA}(\lambda)$, and at 600 nm because that is its usual lasing wavelength. If pumping with a laser (such as frequency-doubled Nd: YAG), then we would consider discrete frequencies for lasing and pumping fluxes. Since we are pumping by flashlamp then we use the approximation for I_P discussed above. The pumping bandwidth is taken as 50 nm, and the useful pump power is within this bandwidth. If efficiency were the only concern, a wider absorption bandwidth would be obtained by increasing the dye concentration. But this would cause uneven heat deposition in the medium. We assume use of a pyrex water jacket to minimize pumping to higher excited states with resultant extra heat deposition. The value for γ_{TO} assumes

Table 6.6. Dye parameters for rhodamine 6G

Parameter	Description	Wavelength [nm]	Value	Reference
σ_{PA}	Pump absorption	530	4.5×10^{-16} cm^2	a
σ_{LA}	Lasing absorption	600	1×10^{-19} cm^2	a
σ_{PE}	Pump emission	530	2×10^{-17} cm^2	a
σ_{LE}	Lasing emission	600	1.3×10^{-16} cm^2	a
σ_{P1}	Pump singlet absorption	530	4×10^{-17} cm^2	a
σ_{L1}	Lasing singlet absorption	600	1×10^{-17} cm^2	a
σ_{PT}	Pump triplet absorption	530	1×10^{-17} cm^2	b
σ_{LT}	Lasing triplet absorption	600	4×10^{-17} cm^2	b
γ_{10}	Singlet to ground spontaneous		2.9×10^8/s	a
γ_{1T}	Singlet to triplet spontaneous		2.0×10^7/s	b
γ_{T0}	Triplet to ground spontaneous		2.0×10^7/s[c]	b
k	Bulk absorption	600	0.002/cm	*
N_0	Population density		1×10^{16}/cm^3	†

a. [6.129]
b. [6.130]
c. Value estimated for equilibrium with air.
* This corresponds to 0.2% loss per cm.
† Based on 2×10^{-5} molar concentration.

triplet quenching by dissolved oxygen. Without this triplet quenching γ_{T0} would be so low that the triplet loss p_{LT} would seriously reduce the efficiency.

From Table 6.6, the values of the various p_{ij}, and the population ratio between the first excited and the ground state are calculated for various combinations of I_L and I_P, and listed in Table 6.7. The estimated quantum efficiency η_L is also calculated from (6.39). The same central wavelength and

Table 6.7. Values of p_{ij} and η_L for rhodamine 6G (using dye values from Table 6.6)

I_P I_L MW/cm^2	10 kW/cm^2			30 kW/cm^2			1 MW/cm^2		
	1.5	4	10	1.5	4	10	1.5	4	10
p_{S1}	0.49	0.18	0.07	0.49	0.18	0.07	0.49	0.18	0.07
p_{ST}	0.03	0.01	0.01	0.03	0.01	0.01	0.03	0.01	0.01
p_{L1}	0.08	0.08	0.08	0.08	0.08	0.08	0.08	0.08	0.08
p_{LT}	0.31	0.31	0.31	0.31	0.31	0.31	0.31	0.31	0.31
p_{L0}	0.04	0.10	0.26	0.01	0.03	0.09	0.00	0.00	0.00
p_{LB}	0.08	0.20	0.51	0.03	0.07	0.17	0.00	0.00	0.00
p_{P1}	0.00	0.00	0.00	0.01	0.00	0.00	0.18	0.07	0.03
p_{PT}	0.00	0.00	0.00	0.00	0.00	0.00	0.05	0.02	0.01
p_{PX}	0.00	0.00	0.00	0.00	0.00	0.00	0.09	0.03	0.01
N_1/N_0	0.014	0.007	0.003	0.04	0.02	0.01	1.2	0.61	0.28
η_L	0.30	0.26	0.05	0.37	0.43	0.36	0.35	0.48	0.55

216   P.N. Everett

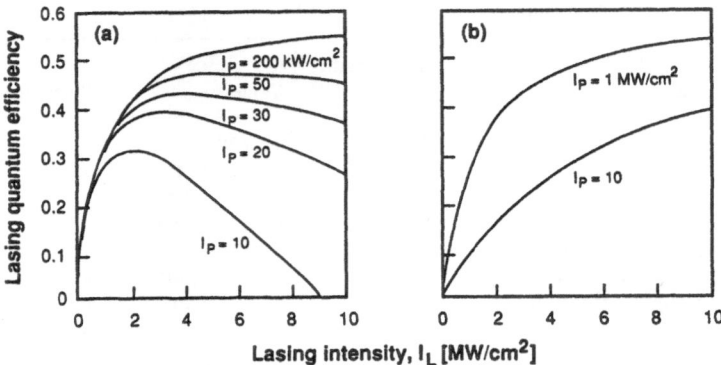

Fig. 6.5. Lasing quantum efficiency vs. lasing intensity for various pump intensities, with absorption coefficient $k = 0.002\,\text{cm}^{-1}$ (a) at pump intensities between 10 and $200\,\text{kW/cm}^2$ applicable to flashlamp pumping, (b) at pump intensities 1 and $10\,\text{MW/cm}^2$ applicable to laser pumping

bandwidth is used for all the pump loss mechanisms. All photons outside that band will be considered lost anyway. Since it is hard to obtain even $30\,\text{kW/cm}^2$ into the 50 nm pump bandwidth, the $1\,\text{MW/cm}^2$ is for laser pumping.

The lasing quantum efficiency is plotted as a function of I_L and I_P in Fig. 6.5. These curves help establish the intensity trade-offs. The importance of having $I_L > I_{\text{sat}}$ (about $1.5\,\text{MW/cm}^2$) is clear. A parameter space for optimal efficiency can be selected from the plot. Increasing I_P will always produce higher efficiency if an associated optimal I_L can be generated. In an amplifier, I_L can be controlled by optically expanding or contracting the beam to be amplified, and in an oscillator by controlling the feedback. The limits on efficiency generally will come from how hard we can pump and how much intensity the optical elements will withstand. The limiting intensities lessen as the pulses lengthen, probably as the inverse square root of the pulse length.

a) Population Distribution and Shifting of Wavelength

For efficiency, the lasing intensity must exceed I_{sat}, which for rhodamine 6G is about $1.5\,\text{MW/cm}^2$. It is difficult to obtain even 2% of this in the pump band from flashlamps so we will generally have $I_L \gg I_P$. Since σ_{PA} and σ_{LE} have approximately the same value then, from (6.33), most of the population will be in the ground state. Ground state absorption will thus push the lasing to longer wavelengths where the absorption will be less. With laser pumping, I_P can be much greater, leading to a higher ratio N_1/N_0 and lower ground state absorption as seen in Table 6.7, resulting in lasing at shorter wavelengths.

b) Dye Degradation

As dye degrades, its absorption increases. The rate at which this causes the output to drop depends on operating conditions. This is illustrated in Fig. 6.6, plotted for rhodamine 6G from the data of Table 6.7. The curves of Fig. 6.6a are for flashlamp pumping, with $I_P = 20\,\text{kW/cm}^2$. Those of Fig. 6.6b are for laser pumping at $I_P = 1\,\text{MW/cm}$. The efficiency falls up to ten times faster in the flashlamp case. These curves should aid understanding when optimizing a system. It may, for instance, be unwise to optimize when the dye is degraded.

The organic dyes used in these lasers tend to degrade when exposed to the UV and near-UV components of the flashlamp output, and create species that absorb at the lasing wavelength and thus contribute to k. Spectral filtering of the flashlamp light can reduce the degrading effect by removing the portion of the spectrum that causes greatest damage to the dye but adds little to the useful pumping. The UV degradation problem was investigated by *Knipe* and *Fletcher* [6.130] and by *Jones* et al. in a series of papers [6.131–136]. It is generally such absorption, and not the loss of dye, that has the major degrading effect on laser output. For instance, the lasers cited earlier [6.74] were equipped with in-line transmission monitors. Laser output dropped 10% when transmission of a 1-meter path of the dye solution, near the lasing wavelength, decreased 5%. Dye loss was insignificant.

c) Triplet Quenching

In laser dyes, spontaneous transitions to the triplet state compete with the lasing transition. Typically the triplet state has a long lifetime and absorbs some of the lasing light, a problem recognized from the earliest days of dye lasers. Increasing the lasing intensity reduces the fraction of molecules going into the triplet state,

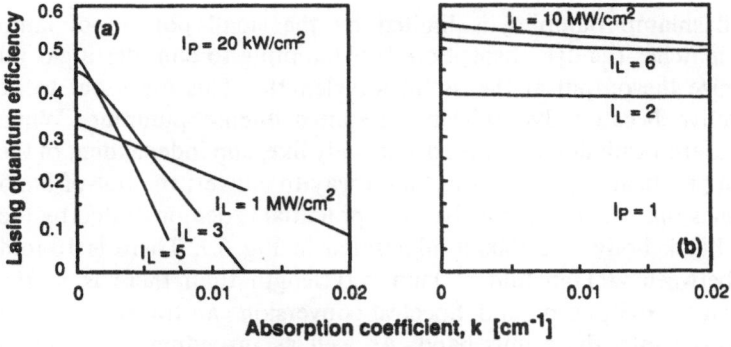

Fig. 6.6. Lasing quantum efficiency vs. absorption coefficient, (a) pump intensity 20 kW/cm² (flashlamp) and lasing intensities 1–5 MW/cm², (b) pump intensity 1 MW/cm² (laser pump) and lasing intensities 2–10 MW/cm²

but this is balanced by the larger amount of lasing light to be absorbed. For pulses of more than 1 μs duration the accumulation of absorbing triplet molecules can terminate the lasing.

Various triplet quenchers have been demonstrated. They exchange energy with the triplets, allowing them to relax to the ground state. The organic quenchers tend to be obnoxious and difficult to use. They also complicate regeneration of the dye solution to recover from degradation. Fortunately oxygen is a good quencher for a number of common dyes, and there is enough naturally occurring in the air that maintenance of air equilibration is frequently sufficient to give marked improvement. In some cases, particularly at high altitude, further improvement can be obtained by enriching the oxygen beyond the air equilibration [6.137]. *Weber* [6.138] reported that nitrogen can also be a triplet quencher with DCM dye, but it is known that it can quench lasing in other dyes by driving out the oxygen.

6.6.10 Shape and Size of Dye Cell and Laser Head

It was shown earlier that the pumped surface area of the dye should approximately match the lamp bore surface area. But the lasing cross sectional area must be much smaller to obtain the high lasing intensity needed for efficiency. This leads to the transverse pumping geometry almost always used, with the length of the dye cell much exceeding its diameter.

The size of the laser head is dominated by the flashlamps, whose size and number will be determined by the energy output required, the pulse length, and the expected efficiency.

6.7 Efficiency Improvement by Spectral Conversion

Efficiency of flashlamp pumping is limited by the small portion of lamp spectrum that matches the dye absorption. It is tempting to consider a gas fill that will optimize the output at the useful wavelengths. This turns out to be counter-productive because dye efficiency requires intense pumping. When driven hard, the lamp emission becomes black-body like, and independent of the fill gas. A more practical approach is to find a way to convert the out-of-band photons into ones that will excite the dye. The potential is demonstrated by the spectrum of a black body at 20 000 K illustrated in Fig. 6.7. There is 10-fold more energy between 200 nm and 500 nm wavelength than there is in the desirable rhodamine 6G pump band. Spectral conversion can transfer some of this wasted energy into the pump band. As well as providing more pump photons, the higher pumping intensity can increase the lasing quantum efficiency.

Wladimiroff demonstrated a number of fluorescent converter solutions for pumping in the visible in 1967 [6.38]. *Peterson* and *Snavely* [6.139] increased the

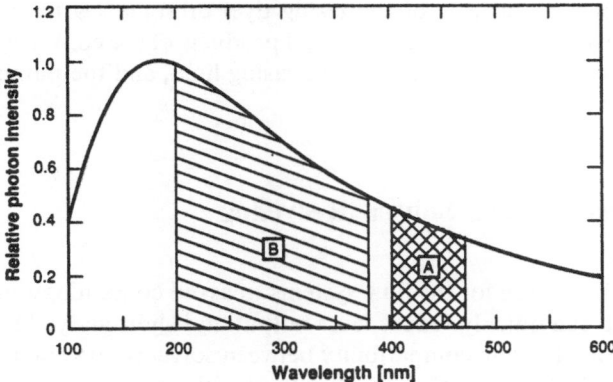

Fig. 6.7. Photon spectrum of a black body at 20 000 K, showing photons (A) within main pump band of rhodamine 6G, and those available for conversion (B)

output of a rhodamine B laser by adding rhodamine 6G to the dye solution in 1968. In 1971, *Moeller* et al. [6.140] used rhodamine 6G as an intermediary dye for lasing cresyl violet using a nitrogen laser pump. *Burlamacchi* and *Cutter* in 1977 [6.141] and *Mazzinghi* et al. in 1981 [6.34] added a film of spectral converter between lamp and dye. *Pavlopoulos* evaluated various dye mixtures in 1978 [6.142]. *Matheson* and *Thorne* in 1979 [6.39] evaluated spectral converters mainly for reducing nonuniform heating of the dye solution. *Rivano* and *Mazzinghi* in 1984 [6.40] developed a model for spectral energy transfer. They experimented with a number of transfer dyes and obtained a twofold increase in laser output. *Fletcher* in 1983 [6.143] and *Lacovara* et al. in 1985 [6.144] demonstrated spectral conversion for solid state lasers. *Everett* et al. in 1986 [6.41] obtained 75% more output from a coumarin 504 laser by adding stilbene 420 to the lamp coolant. *Pacheco* et al. [6.73] reported a laser with a 3-channel dye cell similar to the Burlamacchi and Cutter design, flowing coumarin 420 lasing dye in the central channel, and oligo 373 converter dye in the outer ones, operating at 40 Hz, 3 J per pulse. It is pumped by three lamps each side. The efficiency increased, and less heat was deposited in the lasing dye, which also lasted longer because of less UV. *Lacovara* [6.145] has suggested sodium salicylate as an interesting candidate, either in solution or in the solid form packed into a diffuse reflector, perhaps mixed with barium sulphate. The properties of sodium salicylate have been reported by *Nygaard* [6.146] and by *Kritianpoller* and *Knapp* [6.147]. *Levin* and *Cherkasov* [6.43], in a recent paper, compare various approaches toward spectral conversion.

The converting dye may be mixed with the lasing dye, but this has the following disadvantages. Extra absorption may occur at the lasing wavelength because of excited state absorption or from degradation of the converter dye. The conversion produces heat. If placed in a cell between the lamp and the lasing dye, care should be taken that the converter itself does not lase, perhaps by adding a suitable absorber (e.g. a little of the lasing dye) or some scattering material. If not in the active medium, UV degradation products of the converter dye are less serious because they are not seen by the lasing light, and the pump light passes through only a short path.

6.8 Solvent, Flow Systems, and Solvent Recycling

Materials must be carefully chosen for the flow systems to avoid contamination of the dye solution, which can quickly reduce laser output and shorten dye life. Various tables are available showing compatibility between solvents and engineering materials [6.148–150]. However they must be used with caution, because deterioration of lasing output was probably not a factor in the tabulation. In the author's experience, mainly with acetamide and methanol solvents and coumarin 504 dye, polyethylene is useful for pipes where some flexibility is required. Polypropylene is good where less flexibility is needed and has a wider range of fittings. These plastics are economical and can be fusion welded. Stainless steel is ideal where rigidity and expense are not a problem. It can be welded, and a variety of bolted and compression jointing methods are available. Viton O-rings are acceptable for ethanol, but not for methanol or acetamide, for which Buna-N is preferred.

6.8.1 Flow Through Laser Head

The need for the small cross section for lasing, but large area for coupling of the pump, leads to the typical shapes of the dye cells. It is generally accepted that fresh dye should be introduced into the active volume for each pulse, with perhaps some additional flushing. Longitudinal flow is usually easier to engineer and is essential for coaxial lamps, but for pulse rates much higher than 10 Hz the required flow rate becomes impractical so transverse flow is then used. This leads to a rectangular design for the cell.

The desire to minimize the lasing cross section, maximize the lamp coupling area and minimize the volume of dye passing through to provide fresh dye for each pulse, has led to a number of designs in which the cell has a thin planar shape. It may be less than a few millimeters in width, perhaps a few centimeters in height, and many centimeters long in the lasing direction, with vertical flow.

6.8.2 Regeneration of Dye Solution

Damaged dye solution can be recycled by removing the degradation components. Ideally one would remove only the harmful components, and then replace the small portion of depleted dye. There is very little published on this subject, but Phase-R Corporation supplies alumina filters with some of their commercial lasers, reporting that they clean out degradation products and increase the dye life between two and ten-fold, depending upon the dye [6.75]. A very practical approach is to strip everything out of solution with activated carbon, and then replace the dye. A laser previously cited [6.74] is engineered to flow a variable portion of the dye solution through filters of activated carbon while the lasers are running, to leave only the solvent. Fresh dye is then added to restore the concentration. The dye is added as a concentrated slurry of fine particles which dissolve rapidly in the flowing solution. Candela Laser Corporation and Phase-R Corporation supply capability in some of their commercial lasers for cleaning the solution with activated carbon between lasing operations. Cleaning the solvent and adding more dye will maintain the laser output, and remove the problem of disposing of large quantities of spent dye solution. The technology is relatively simple and inexpensive. The filters and housings have been developed for cleaning water on a large scale and are readily obtainable.

An advantage of a built-in cleaning system is that the initial charge of solvent need not be as pure, since it can be cleaned by circulating through the carbon before any dye is added. This is important in large systems since it permits a less expensive grade of solvent and relaxes the cleanliness requirements on shipping containers. It also finishes the initial cleaning of the dye handling system.

6.8.3 Optical Quality of Medium and Other Factors

These lasers inherently have long lasing paths through the active medium, and so are very sensitive to the optical qualities of the dye solution. Any variations in refractive index will cause phase distortions in the lasing beam. The flow system must flow the dye through the active cell without disturbing its optical quality. The optical pumping must cause minimal perturbation. Optical quality of the medium may be important for spectral stability [6.151] and beam quality.

Refractive index perturbations result from temperature and pressure variations. Sudden temperature variations lead to acoustic waves and hence to further index variations that develop with time. At the start of pumping, optical quality is that of the passively flowing dye solution. As heat is deposited, the temperature rises and the refractive index changes. If the heating is uniform then there is initially no optical distortion because the temperature rises uniformly. However, the resultant expansion of the dye solution causes acoustic waves which reflect off the walls of the cell. The consequent pressure variations then cause non-uniform refractive index and hence phase distortion in the lasing beam. The non-uniformities propagate at the velocity of sound, which is about 1 mm/μs in liquids. Thus, depending upon the size of the cell, one can estimate how long it takes for the acoustics to become important.

The thermal refractive index coefficient at constant volume is important at the beginning of the pumping. Later, when the molecules have had time to move (limited by the sound velocity), the thermal coefficient at constant pressure becomes important. For pulses lasting much less than 1 μs there is insufficient time for the acoustics to propagate before the lasing is over. Pulses lasting much more than 1 μs are dominated by the acoustics for most of the pulse. Modeling of such pulses is difficult because they start in one regime and end in another.

It is generally considered that dye solution should be pumped through the laser heads fast enough to ensure at least one interchange between pulses, that temperature fluctuations in the passively flowing liquid should be kept within ± 0.1 K, and bubbles and scattering particles should be eliminated. This requires careful design of the flow system. The design of such a system is outlined in [6.74]. Water is a favored solvent because of its high heat capacity. However, many dyes do not perform well in water, often because of dimerization or lack of solubility, so organic solvents are often used, sometimes mixed with water. Often there is a trade-off between maximizing water content for optical quality and organic solvent for higher output. Flammability may also be a concern pushing for a higher water content, or for a non-flammable organic solvent such as acetamide.

In a well-designed flow system the optical distortions come mostly from heat deposited by the flashlamp pumping. Some solvents, such as ethanol and water, have strong IR absorption bands [6.16]. *Gavronskaya* et al. in 1977 [6.152] used holography to investigate the phenomena. As many as fifteen fringes of distortion were observed when using an alcohol solution, but only two with aqueous solutions. Acoustic perturbations were still observable 300 μs after the pulse. *Aristov* et al. [6.153] did further holographic studies in 1978. They found that acoustic disturbances from the lamps and intervening filters could also be important if not decoupled. They confirmed that optical distortion arises mainly from absorption of pump power when the pump duration equals or exceeds the time for propagation of sound across the dye cell.

6.9 Resonators and Propagation

Stable resonators were generally used in the early days. In these the phase and amplitude distribution repeat on successive passes, and the output is generally extracted through a partially reflecting mirror. To give good beam quality, the Fresnel number of a stable resonator [6.2] should not exceed unity, otherwise multiple transverse modes can be sustained. When significant power is required, a Fresnel number of unity in a stable resonator leads to either an impractically low medium diameter, or to undesirable length. Resonators with flat mirrors are theoretically on the borderline between stable and unstable, but in practice become one or the other from lensing in the medium during the pumping. In unstable resonators the phase and amplitude distribution grows larger on successive passes, and the output is extracted from the outer portion of the resonator. Higher order and off-axis modes expand faster and see the gain for a lesser number of passes, and hence are discriminated against, even with high Fresnel numbers. This allows a larger diameter of active medium to be used and is often more practical for high power operation.

Waveguide resonators are attractive, but they lead to cross sectional areas too small for high power. A hybrid, consisting of a waveguide in one transverse dimension, and stable or unstable in the other, has proven practical. The output is then narrow in one dimension, but the aspect ratio may be improved with cylindrical optics. *Burlamacchi* et al. and *Mazzinghi* et al. have reported on such lasers [6.115, 34, 36].

Resonators are discussed by *Siegman* [6.2] and by *Steier* [6.154]. We will explore some topics of particular interest, starting with basic concepts. Unstable resonators have proven particularly useful in high power FEDLs and so will be examined more closely.

6.9.1 Concepts

The geometric approximation ignores diffraction, and "perfect" beams are considered to have purely spherical wave fronts, appearing to emerge from, or head towards, a point. The "geometric divergence" is the reciprocal of the radius of curvature of the wave front. When the geometric divergence is zero, then the beam is collimated and the wave front is plane. Positive geometrical divergence will indicate an expanding beam. Many of the properties of unstable resonators can be developed from geometric optics.

Beams are never geometrically perfect in practice. Diffusion, diffraction, distortion, small-angle scattering, etc., always adds some "angular divergence." This can be described by the light emanating into a cone from any portion of the spherical wave surface. Hence, any practical beam has a geometric divergence (positive, negative, or zero) plus some angular divergence superposed (inherently positive). This is a simplified picture but will carry us quite far before we get into trouble.

A useful concept is that of brightness [6.155–157], less popular than in earlier days, but very useful in obtaining insight on propagation of light in optical systems. It is a concept that becomes better understood and appreciated as it is applied to more cases. Its application can save much tedious optical analysis. Brightness is defined by

$$B = \frac{P}{A\Omega n^2} \, [W\,cm^{-2}sr^{-1}], \tag{6.43}$$

where P, A and Ω are incremental quantities of power, cross sectional area normal to the propagation, and solid angle containing P, and n is the refractive index in the medium. The solid angle Ω comes from the "angular divergence" of the earlier discussion, not the "geometric divergence." Spectral brightness is the derivative of B with frequency or wavelength. Brightness applies to surfaces and to propagating beams. The brightness of a black-body surface is the same for all angles of observation, and is determined by the surface temperature via the black-body laws of thermodynamics. The brightness of laser sources and propagating beams can be very dependent upon the observation angle.

The "brightness theorem" states that brightness is, at best, conserved when propagating through a passive optical system. It follows from the laws of thermodynamics and the relationship of brightness with black-body temperature. For example, when a beam is expanded in a lossless passive system, the brightness theorem tells us that the angular divergence decreases proportionately as the diameter increases. This is why large diameter telescopes are needed for transmitting beams over long distances. Scattering or absorbing losses, and some optical distortions, will reduce brightness. In the geometric limit, with no angular divergence, the brightness would be infinite, which is physically impossible since it implies a source with infinite equivalent black-body temperature. To some extent the concept of radiance has replaced that of brightness in modern day literature, at some loss, because radiance is commonly limited to emitting surfaces.

Related concepts are those of étendue used in optical instruments [6.114], and the solid angle associated with a single mode [Ref. 6.158, p. 548]. If a black body is emitting through an aperture of area A, into a medium with index n, then the solid angle associated with a single mode is $\Omega = \lambda^2/An^2$. This allows estimation of the solid angle into which a multi-mode laser is propagating, and thence its brightness.

6.9.2 Unstable Resonators

A full analysis of an unstable resonator uses computer code to follow the electromagnetic fields as they propagate back and forth. It is complicated for dye lasers with pulses of 1 µs or longer because of the time-varying refractive index variations. Fortunately, a simpler analysis can help understanding and design, although it does not give details of the mode structure. We will first use the geometric "self consistent" approach to determine quasi steady-state beam characteristics, and to estimate sensitivity to misadjustment and lensing. We will then use the geometric "virtual source" approach to understand initial development of the beam. We will then introduce Gaussian propagation to bring in diffraction. This will give further insight, and help understand aligning techniques that rely on directing light backwards into the resonator. The geometric approximation does not allow such a beam to emerge again.

a) Geometric "Self-Consistent" Analysis

A generalized unstable resonator is shown in Fig. 6.8. Spherical mirrors M1 and M2 form the ends of the resonator, with radii of curvature R_1 and R_2, respectively. The focal lengths are equal to half the radii of curvature. High power lasers often use a confocal geometry, to obtain a collimated output, with the common focus outside the resonator to avoid energy concentration. This requires a convex output coupler and concave back reflector, separated by the difference in magnitude of the mirror focal lengths. The output is generally coupled out around the output mirror, or the same effect achieved with an

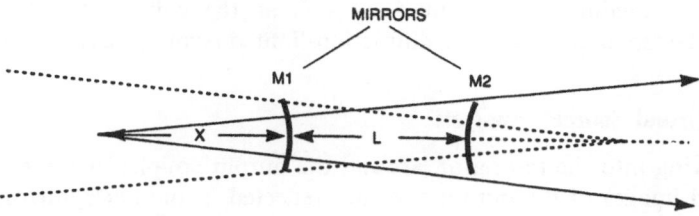

Fig. 6.8. Unstable resonator geometry

internal "scraping" mirror with an elliptical hole, placed at an angle inside the resonator. Hence the output usually has a hole in the middle. The hole can be off-center, or even on an edge, using off-axis optics. The general self-consistent geometric solution [6.159] is an output beam that appears to geometrically diverge from a distance X behind the rear mirror, given by

$$X = \frac{(\pm \sqrt{1 - 1/g_1 g_2} - 1 + 1/g_1)L}{2 - 1/g_1 - 1/g_2} ,$$

(6.44)

where

$$g_1 = 1 - \frac{L}{R_1}$$

(6.45)

$$g_2 = 1 - \frac{L}{R_2} .$$

(6.46)

In these relations, L is the resonator length, and R_1 and R_2 are the curvatures of the rear and output mirrors, respectively. A focusing (i.e. concave) mirror has positive radius of curvature. If X is negative, with magnitude exceeding L, then the output will converge.

b) Example

We will use the resonators in an operational laser system [6.74] as an example. The confocal resonator have values of 2.5, 8 and -3 m for L, R_1 and R_2. Equations (6.44–46) give an infinite value for X, so a collimated output is expected. In practice the output is not quite collimated because of imprecision in the radii, placing of the reflectors and thermal lensing in the medium.

Sensitivities to errors in L, R_1 and R_2 are found by varying the values one at a time. In the application, more than 0.025 m^{-1} of divergence from the lasers, i.e. about $10 \times$ the diffractive full angle, reduces the usefulness. Analysis shows that this geometric divergence would be caused by an error of -35 cm in L, or 70 cm in either R_1 or R_2. It would also cause a beam matching the 14 mm output aperture to be only 13.2 mm diameter at the back end of the laser. Thus efficient use of the medium would suffer, as well as the collimation. The sensitivity would be ten times greater if a diffraction-limited beam were required.

c) Geometric "Virtual Source" Analysis

An observer looking into the resonator, around the output coupler, will see a series of repeated images of the output aperture reflected in the back mirror. They will be different sizes, at different distances, corresponding to the multiple internal reflections in the resonator [Ref. 6.2, p. 894]. At the beginning of the pulse only the nearer virtual sources will be bright. As the pulse progresses, the further ones will become bright. After a time equivalent to N round trips, the brightest sources will be those at distances corresponding to approximately N internal reflections from the back mirror. It is a simple matter to set up a computer program to find the diameters and distances of these successive virtual sources, using conventional imaging analysis. The results for the design values of the example resonator are shown in Table 6.8.

Table 6.8 shows that the closer sources appear larger, giving rise to a beam with high angular divergence, but beyond the fifth, the sources appear many kilometers away and unresolved through the output aperture. Hence, after only 100 ns this geometric analysis says the beam will be collimated within the diffractive limit, if no distorting phenomena intervene. In practice the beam is found to be about twenty times the diffractive limit, indicating the distortions do occur. These may come from refractive index fluctuation in the turbulent medium enhanced by heat from the flashlamp pumping, and interaction with phonons, tiny bubbles or molecular scattering processes.

d) Sensitivity to Errors in Resonator and Lensing in Medium

The same "virtual source" approach may be used to find sensitivities to resonator length and mirror curvatures. When this is done, the results satisfyingly converge to those of the previous "self-consistent" geometric solution

Table 6.8. Example resonator. Virtual source size and distance as pulse progresses

Reflections	Time	Virtual source Distance	Diameter	Divergence Geometric	Angular
1	17 ns	6.7 m	0.037 m	0.15 m^{-1}	5.5 mrad
2	33 ns	71.9 m	0.010 m	0.014 m^{-1}	1.4 mrad
3	50 ns	535 m	0.27 m	1.87 km^{-1}	0.50 mrad
4	66 ns	3.83 m	0.71 m	0.26 km^{-1}	0.18 mrad
5	83 ns	27.3 m	1.9 m	0.037 km^{-1}	69 µrad
6	100 ns	194 m	5.0 m	0.005 km^{-1}	26 µrad
7	120 ns	1400 km	13.7 m	0.0007 km^{-1}	9.7 µrad

Geometric divergence = 1/distance.
Angular divergence = source diameter/distance.

Table 6.9. Sensitivities to imperfections of resonator. The errors listed result in positive geometric divergence equivalent to about $10 \times$ diffraction.

Item	Nominal value	Error
Resonator length	2.5 m	−35 cm
Back mirror radius	+8.0 m	+70 cm (−0.022 dioptre)
Front mirror radius	−3.0 m	+70 cm (−0.14 dioptre)
Lensing (each head)	0.0 dioptre	−0.01 dioptre

Notes 1. Lensing of as much as 0.04 dioptre per head will cause an oscillatory beam.
      2. Lensing "power" of a lens = $1/f$ dioptres, where f is expressed in meters.

after the first half dozen round trips. The same "virtual source" approach finds sensitivity to medium lensing. In the same example, when two lenses are placed $L/4$ from the ends of the resonator to represent the laser heads, then resultant sensitivity is $30 \, \mathrm{m}^{-1}$ of geometric divergence per dioptre of each lens. (The focusing power of a lens, in dioptres, is the reciprocal of the focal length in meters.) If both lenses are superposed in the middle, the result is the same, so exact lens position is not critical. The sensitivity is quite linear for negative lensing to beyond -0.1 dioptre. However, at $+0.04$ dioptre the solution starts oscillating between positive and negative divergence. This is attributed to crossing into the stable resonator regime in which it is known that a geometric solution will break down. In that regime we would expect the solution to be a much smaller beam, which may result in mirror burning. This is of concern, because the lensing is expected to be positive due to stronger heat absorption at the outer edges than at the center. The collected sensitivities are summarized in Table 6.9.

e) Alignment of Unstable Resonators

Unstable resonators may be aligned by injecting a collimated cw laser beam back into the resonator [6.160, 161]. The injected beam may be small enough to fit within the annular opening around the output coupler, but the author has found it more sensitive to inject a collimated beam the full diameter of the laser medium. The light enters through the output annulus, then collapses on successive passes through the resonator, down to the diffractive core. It then expands again, and after a like number of further passes it emerges through the output annulus. Provided the alignment is good, it closely simulates the actual pulsed laser beam. This gives sensitive alignment, since very slight misadjustments of the resonator mirrors are apparent when the cw output beam is observed. This cw output beam is also useful for aligning optical elements outside the resonator, for accepting the laser's pulsed output.

6.9.3 Gaussian Analysis of Unstable Resonators

Much of the resonator performance can be understood from geometric optics, but not the above aligning procedure. Geometric optics says that a collimated beam projected into a correctly aligned confocal laser will be trapped in a beam, forever becoming smaller. However, the *Kogelnik* and *Li* [6.162] approach for projecting Gaussian beams through optical systems forecasts the observed behavior and gives useful insight. A lowest-order Gaussian beam can be represented in any plane by a complex parameter q, given by

$$q = \frac{1}{R} - \frac{i\lambda}{\pi\omega^2} , \tag{6.47}$$

where $\omega(z)$ is the spot radius at which the intensity has dropped to $1/e^2$, and R is the radius of curvature of the wave front. When traveling a distance or passing through an optical element, q_1 becomes modified to q_2 where

$$q_2 = \frac{Aq_1 + B}{Cq_1 + D} . \tag{6.48}$$

Initially $A = D = 1$ and $B = C = 0$, but they go to new values on each successive optical operation. When propagating over distance z, with no intervening optics

$$A \to A + Cz, B \to B + Dz, \quad C \text{ and } D \text{ are unchanged} . \tag{6.49}$$

When passing through a thin lens with focal length f

$$C \to C - A/f, D \to D - B/f, \quad A \text{ and } B \text{ are unchanged} . \tag{6.50}$$

The spot radius of a beam at collimation is ω_0. A useful parameter is the "Rayleigh distance" z_R defined by

$$z_R = \frac{\pi\omega_0^2}{\lambda} . \tag{6.51}$$

At this distance from the collimated waist the spot radius has increased by the ratio $\sqrt{2}$, and the central intensity has halved. The half angle subtended by a beam in the far field is

$$\Delta\theta_f = \frac{\lambda}{\pi\omega_0} . \tag{6.52}$$

Higher-order Gaussians have their intensity pushed out to larger diameters and are more difficult to handle. However, the pattern associated with any order maintains itself as it propagates, and maintains the same scaling to the lowest order Gaussian. The conclusions from the next analysis will apply insofar as a Gaussian beam is a good model of the definitely non-Gaussian ones used for the aligning, and resulting from the lasing. It will be shown later that essentially the same results are obtained when an arbitrary intensity distribution is considered as a sum of higher-order Gaussians.

It is instructive to set up a simple program on a personal computer, using the above relations, to propagate a lowest-order Gaussian beam through the successive optics of a confocal resonator. It will confirm the initial collapsing to the diffractive core, followed by expansion to emerge collimated. It is easy to investigate the effects of errors in the resonator and lensing in the medium. The resonator "solution" is found by adjusting the radius of curvature of the ingoing beam to obtain an exciting beam whose emerging divergence is matched to the entering convergence. The sensitivities to errors and lensing agree with the previous analyses. The same is found for resonators that are not nominally confocal, and indeed for stable resonators. This appears to give a good model of the operation of lasers in general, and a feeling of insight is gained after running a few examples [6.163].

In summary, it is found that a collimated alignment beam injected directly into a correctly adjusted confocal resonator will indeed emerge collimated, providing a good simulation of the pulsed output beam. If the alignment beam is not collimated, then the emerging beam will be different from the entering beam, and diagnostics will show it focusing in a different plane after passing through a lens.

a) Higher-Order Gaussian Analysis

Introducing higher-order Gaussians allows better matching of a realistic energy distribution, such as the annular one from the laser. We have mentioned that a higher-order Gaussian maintains its particular intensity distribution, varying only the scale factor as it propagates (as does the lowest order). The patterns of all orders of the same fundamental grow or shrink in the same proportion. This means that if we can relate a significant radius ω_{mn} in a higher-order Gaussian to the radius ω_p of its lowest order in one plane, then the ratio (ω_{mn}/ω_p) will remain constant as they propagate.

In the example, the lowest-order solution is considered to have a waist radius of 7 mm as it enters or leaves the laser. Now let us consider a smaller lowest-order beam only one quarter of this size, i.e. spot radius of 1.75 mm. Let us suppose we can generate higher-order Gaussians based on this fundamental which, when summed together, will give a better approximation to the annular lasing or injected beam. We will refer to our new composite beam as the "4 × Gaussian" beam. Since we have chosen it to approximately match the laser's input aperture, then its outer diameter will be 7 mm as it enters the laser. Thereafter its propagation will follow that of the lowest order, except scaled to a 4 × larger pattern in every plane. Hence, the same computer program will propagate it, provided we launch it into the resonator as a lowest-order Gaussian with 1.75 mm spot radius. It will completely exit again when the spot radius of the lowest-order Gaussian expands once again to 1.75 mm. Following the above rules, its behavior in the resonator is found similar to that of the lowest order excepting that the beam will exit after a smaller number of trips, and the typical "solution" divergences are greater.

Now let us consider the far-field. Analysis of the example laser shows that the 4 × Gaussian has a far-field size 16-fold greater than the original one. One factor of four has come from the lowest order being smaller, and the other from the 4 × scale expansion because of the higher order. We have thus added a way of handling far-fields more like those observed in practice.

6.9.4 Waveguide Resonators

Waveguide lasers have been reviewed by *Degnan* [6.164]. The refractive index of the medium is generally less than that of the dye cell, so energy is not contained by total internal reflection as it would be in conventional dielectric waveguides. However, the glancing incidence Fresnel reflections at the cell walls give relatively small losses for low-order modes when the transverse dimensions are large compared with the wavelength, but small enough that not too many modes are allowed. With flashlamp pumping, the waveguide is only practical in one dimension. If it were in both dimensions, the coupling from the lamp would become difficult, and only very low powers would be practical because of the small cross sectional area. Hence, the other dimension is run in a non-waveguide configuration. *Burlamacchi* et al. and *Mazzinghi* et al. have reported on a number of such lasers [6.115, 34, 36], typically about 0.5 mm thick in the waveguide dimension, and a few cm wide in the other transverse direction. The resonator mirror optics are cylindrical, to match the rapidly diverging wave-guide mode external to the dye cell from the waveguide dimension, and have no optical power in the other dimension. They achieved as much as 5 J/pulse in 15 μs pulses at 10 Hz pulse rate using three linear 6-inch lamps close-coupled to either side of the cell. The output is strongly divergent in the waveguide dimension. The author and others, in unpublished work, have used an oscillator and amplifier configuration, both waveguided in one dimension, spatially fil-tered to obtain 100 mJ pulses with spectra controlled to within 0.5 GHz, in a 4 × diffraction limited beam in both dimensions. The optics used commercial cylindrical eye glass lenses for independently controlling the propagation in the two transverse dimensions.

6.9.5 Propagation

Any propagation over distances approaching or exceeding the Rayleigh dis-tance, D^2/λ, must take diffraction into account (D is the transmitting aperture diameter and λ the wavelength). The reader is referred to [6.2] for the full diffractive treatment. For those who are not mathematically minded, much of the physics and engineering is obtained in the following brief discussion, which allows at least approximate conclusions to be drawn in most situations.

The Heisenberg uncertainty principle describes an inherent fuzziness that must exist in any attempt to describe nature [Ref. 6.165, pp. 6–10]. It is fundamental to the understanding of physical optics. One expression of it says that, for a photon, the minimum uncertainty in its position $\pm \Delta x$, and in its momentum $\pm \Delta p_x$ are related as

$$\Delta p_x \Delta x \geqslant kh \ , \tag{6.53}$$

where h is Planck's universal constant. The value of k is variously described as unity [Ref. 6.165, pp. 5–10, 383], as $1/(2\pi)$ [Ref. 6.165, pp. 2–6 and Ref. 6.166], and as $1/4\pi$ [Ref. 6.158, pp. 1, 15]. Such differences arise from how uncertainty is precisely defined (if this is not a contradiction in terms). However, setting such detail aside, when a photon passes through an aperture, its uncertainty in transverse position is bounded by the size of the aperture. This results in a finite uncertainty in the transverse component of the momentum. The full momentum is h/λ. If the photon is nominally traveling in the z direction, and $\Delta \theta_x$ is the angular uncertainty into the x direction, then $\Delta p_x = \Delta \theta_x h/\lambda$. This, from (6.53), leads to uncertainty in the precise angular direction $\Delta \theta$ as it leaves the plane of the aperture, given by

$$\Delta \theta_x = \frac{k'\lambda}{D} \ , \tag{6.54}$$

with the precise value of k' depending upon the details of the light distribution within the aperture and how the angular uncertainty $\Delta \theta$ is defined. For a square or circular aperture, if we define $\Delta \theta_x$ as the half angle of a cone containing half the power in the far field [Ref. 6.2, p. 740], then $k' \approx 0.4$. Hence the half angle in the far field containing half the power is, in both cases, approximately $0.4\lambda/D$.

In summary, when a collimated beam of reasonably uniform power distribution passes through an aperture of dimension D, then at distances much less than the Rayleigh distance D^2/λ, the gross power distribution remains almost unchanged. At distances much greater than the Rayleigh distance, the gross power distribution expands in a conical angle with apparent origin close to the aperture. The half-conical angle containing half the energy for square or circular apertures will approximate $0.4\lambda/D$. For further details, including evaluation of cone angles for different proportions of the power, the reader is referred to [Ref. 6.2, p. 740] and [6.167].

A sometimes useful approximation is to consider a laser beam to have a Gaussian distribution. The field pattern along the entire Gaussian beam is then characterized entirely by the single parameter ω_0 at the beam waist, plus the wavelength λ [Ref. 6.2, p. 665]. The waist radius $\omega(z)$, containing $1/e^2$ of the total power at distance z, is

$$\omega(z) = \omega_0 \left[1 + \left(\frac{z}{z_R} \right)^2 \right]^{\frac{1}{2}} \ , \tag{6.55}$$

where the Rayleigh distance z_R is defined as

$$z_R = \frac{\pi \omega_0^2}{\lambda} \ . \tag{6.56}$$

At distances much greater than the Rayleigh distance, $1/e^2$ of the total power will be contained in a cone of half angle $\lambda/\pi\omega_0$, i.e. $0.32\lambda/\omega_0$. High power FEDLs will generally expand faster than this, unless the distortions in the dye medium have been brought under control.

a) Diffraction Close to an Aperture

We have discussed the gross behavior of the power distribution far downstream of an aperture. There are also some detailed effects, relatively close to the aperture, that should be considered. When a beam of coherent light is interrupted by an edge, then diffraction effects near the edge of the beam appear quite a short distance downstream. The reader is referred to [6.2] for exact analysis. Approximate solutions can be obtained for individual situations using the graphical Cornu spiral approach [Ref. 6.165, pp. 30–9]. Within the beam that passes, ripples of intensity occur near the edge. These fringes are already well developed at 5% of the Rayleigh distance z_R, and have a dimensional scale of approximately D/N, where D is the dimension of the aperture and $N = z_R/z$. The intensity at the projection of the sharp edge quickly reduces to 25% of the uninterrupted intensity. As the beam propagates, the ripples extend further toward the center. The intensities reached in the peaks of these edge ripples can exceed the mean intensity of the uninterrupted beam by about 25%. This can damage optics if the energy density were already approaching damage limit. Photographs of the ripples have been published by *Jenkins* and *White* [Ref. 6.166, p. 396]. These edge ripples appear whenever there is a hard-edged aperture. If the aperture has a circular shape, the strong symmetry also causes pronounced on-axis ripples of much greater magnitude, and of the same approximate dimensional scale. These also tend to be well developed even at 5% of the Rayleigh distance, with on-axis intensity fluctuating between zero and *four times* the average intensity. They can easily burn optics in high power lasers. *Siegman's* book [6.2] has instructive illustrations of sample near-field intensity patterns for rectangular apertures (p. 721) and circular apertures (p. 730).

Surprising effects can also occur downstream of a central obstacle as in the case of an annular beam, which is typical from an unstable resonator. With almost any shape obscuration, an intense spot will generally occur on-axis after the obscuration. It is variously called the "Poisson spot" or the "spot of Arago" [Ref. 6.2, p. 736]. Although such a spot contains only a minute portion of the total energy, it can have four times the intensity of the major beam, and has caused unwelcome burn spots both inside and outside unstable resonators. Various attempts have been made to reduce the local peaks of intensity by "softening" apertures and obstacles, or by having jagged edges. Often their attractiveness is diminished by other trade-offs, such as inefficient use of some of the medium.

6.9.6 Optical Elements

The resonator mirrors for high-power FEDLs must typically withstand more than 1 MW/cm^2 peak intensity. In most applications the peak intensity is more damaging than the average power. Selection of optics is becoming easier as more suppliers are willing to specify damage limits.

Multilayer dielectric coatings are normally used for mirrors on quartz or high-quality glass. Successive batches from the same supplier may vary in damage resistance. A good mirror should absorb less than 0.2% of the incident laser light. The best mirrors appear to withstand about 10 J/cm^2 in a pulse lasting a few microseconds. In the absence of better information, one can assume that the pulse energy limit is proportional to the square root of the pulse length in this regime.

An infrared thermal camera will give sensitive testing of mirror absorption, down to 10^{-5}, using a technique published by *Draggoo* et al. [6.168]. The method can be used on optics while performing in a high-power laser, and will reveal absorbing inclusions as well as the background absorption.

6.9.7 Isolation

With a high-gain amplifier, a small portion of the output, if fed back, can result in disastrous intensity at the input. In general, the output optics will be working near the damage limit because of efficiency considerations. Hence, if the gain were 10^6, feedback of even 10^{-5} would severely overstress the input optics. The situation is even worse if the amplifier chain has a smaller cross section at the input, as would be optimal in a multi-amplifier system.

An otherwise satisfactory oscillator and amplifier chain may suffer burning optics in the oscillator if the output is focused for an application. The 'cats eye' effect, in which light reflected from a focal plane is retro-reflected, can give strong feedback (as anybody who has survived encountering a tiger at night while using a flashlight will testify). Measurements of far-field often involve placing a sensor in a focal plane and can result in such undesirable feedback.

Various isolators can be used to prevent feedback. They include Faraday rotators, polarizers and quarter wave plates, and attenuators. If the pulse is short, sufficient distance can ensure that the gain cycle is complete before the feedback returns, but with pulse lengths measured in microseconds the distances can become rather long.

In many cases the application does not require focusing of the laser light. If care is then taken to avoid reflections off optical components (by suitable angling), isolation may not be needed. Often the greatest danger is from monitoring equipment, but this can usually be handled by using sufficient attenuation in the monitoring path to ensure that the two passes necessary to return to the laser will produce sufficient attenuation. A simple approach to obtain such isolation in a monitoring path is to tap off some of the main beam with a wedged trasmitting flat with one surface AR coated. If the Fresnel reflection is taken at nearly normal incidence, then polarization effects are eliminated and the reflection can be obtained accurately from knowledge of the refractive index. If the reflected light is then directed at near normal incidence onto another similar optic, not AR coated, then a succession of beams is generated from the multiple reflections, each with a precisely determined attenuation. The wedging of the optics causes the various beams to separate and so eliminates interference effects. The approach is attractive since it uses only relatively inexpensive components that readily withstand the required intensities. Good quality is needed so as not to distort the beam. Five reflections from a glass–air interface will give attenuation of about 10^{-7}. Care in design and use of systems will often eliminate need for the more exotic isolators. If needed, there is generic literature on them which will not be detailed here.

6.10 Pulse Length

The shortest pulses come from mode locking, when all the resonator modes are briefly in phase at the same instant. The result is a short pulse circulating back and forth around the oscillator, with a portion leaving through the output mirror once each round trip. Ideally, the inverse spectral bandwidth determines the pulse length. For a dye laser the bandwidth can be 100 THz, which should give 10 fs pulse length. In practice, dispersion and non-linear effects lengthen the pulse [6.169]. Mode locking requires transfer of energy between resonator modes, either by active modulation of a resonator parameter or by a passive non-linear effect, such as from an absorbing dye.

Passive mode-locking of FEDLs was demonstrated by *Schmidt* and *Schäfer* [6.18], using cyanine dye as a saturable absorber, with rhodamine 6G as the lasing medium. They obtained pulses shorter than their instrument resolution of 0.4 ns. More recently *Singh* [6.20] has reported pulse lengths of less than 1.5 ps using rhodamine 6G and the saturable dye DODCI, in a ring oscillator. However, the shorter pulses only appeared near the end of the train of pulses. Results of various investigators are summarized by *Sibbett* and *Taylor* [6.170], who achieved pulses of less than 3 ps, in the wavelength range 552–583 nm, using rhodamine 110 dye. A recent full discussion of mode-locking in dye lasers is given by *Diels* in [6.8], but the emphasis is on laser-pumped lasers.

When the upper state of a lasing transition has a long lifetime, energy can be stored in the inverted population, and then released quickly by manipulating the optical feedback, or Q, of the cavity. However, dyes typically have only a few nanoseconds lifetime for the upper state, so such Q-switching is not practical. However a related process, "cavity dumping", has been used to store energy in the optical field in a resonator, and then switch it out in a pulse which lasts for one transit time of the resonator. *Morton* et al. [6.21] obtained 0.3 J in a 20 ns pulse by this technique, with energy efficiency 85% of that when the laser was pulsed normally to obtain a 0.3 μs pulse. The peak power was 10-fold higher in the 20 ns pulse. A Pockels cell and polarizer were used for the switching. Pulse shapes have also been controlled by including electro-optic elements, as in the "NESSY" supplied by Phase-R Corporation to the US Naval Research Laboratory in 1978 [6.75], as an illuminator for underwater photography. This laser gave 1 J in a rectangular pulse with 0.35 μs width, and could also be cavity dumped to give 20 ns pulses.

Without special efforts, the pulse length is determined by the flashlamp pulse length, provided the pulse is not terminated prematurely by triplet or thermal effects. Pulses of more than a few microseconds duration, with good beam quality and high efficiency, require careful engineering because of thermal effects in the medium, power density limitations of the lamps, optical window burning, and triplet accumulation. These problems have been attacked by *Janes* et al. [6.17], using FEDLs to pump a dye cell with a "sweeping" technique. A rotating mirror is included in the resonator. It synchronously sweeps one or more FEDL pump beams, and the axis of the resonator, across the face of the dye cell at supersonic speed, while maintaining a fixed direction for the incoming pump beam, the remainder of the resonator and the outgoing laser beam. The small scanned spot allows high pumping intensity (2 MW/cm^2), and is swept across new dye during the extended pulse, leaving disturbed medium behind. The pump laser need not have high beam quality, and a number of them can be time-multiplexed into the pumping. A 100 μs pulse is practical, with exposure of only 0.3 μs for any portion of the dye. Energy conversions in the swept dye cell of about 40% were demonstrated, with expectation of $4\times$ diffractive limited output beam in a 100 μs pulse. Coumarin 504 was used in the FEDL to pump rhodamine 690 in the dye cell. *Neister* also reports achieving 400 μs pulses containing 5.5 J at 0.27% efficiency using coaxial lamp excitation [6.75].

6.11 Spectral Control

Tuning and bandwidth control can be with etalons, prisms, gratings, or birefringent filters in the resonator. It is relatively easy to obtain bandwidths of about 1 nm, even in a high-power oscillator, by using a prism or a birefringent filter. But for narrower spectra the extra elements needed in the resonator lead to loss. Unstable resonators are attractive for high power, but present particular problems in a narrow spectral tuning because of the inherent angular diversity of the lasing light. For instance, in a confocal resonator the emerging light is collimated, but on the previous pass it is geometrically diverging. Birefringent filters can be used in unstable resonators [6.74] to obtain spectral width at least as small as 1 nm. For much narrower bandwidths it may be more practical to generate the spectrum in a low-power stable oscillator and follow it with a power amplifier.

Unfortunately, amplifiers are efficient only when highly saturated. Hence, for high gain, many amplifiers may be necessary. Alternatively, many passes through a single amplifier may be used, but this becomes more complex and there is the danger of cross-talk between the passes. Perhaps the simplest multiple-pass amplifier is one which looks much like a confocal unstable resonator, consisting of a concave rear mirror and a convex output mirror to give an annular output. The signal to be amplified is introduced and collimated through a hole in the concave rear mirror. It then bounces back and forth, expanding on each pass, until it is large enough to exit around the convex output coupler. The size of the entering beam should be large enough that there is not diffractive coupling between successive passes.

Low-power oscillators are discussed in other chapters. The same techniques may be used with FEDL oscillators, so we will confine this discussion to techniques that have proven particularly useful. Spectral width down to a few megahertz can be obtained with multiple air-spaced Fabry-Pérot etalons [6.46], but the losses lead to low efficiency. Multiple solid etalons, tuned by tilting, are less selective but lead to higher efficiency, especially if uncoated. The author and co-workers have used up to three solid uncoated etalons, and a diffraction grating at grazing incidence to couple out of the resonator, to obtain band-widths controllable down to 0.5 GHz.

Various approaches have been used to obtain spectral agility, including switching pulses from different oscillators into a single amplifier and switching dyes while the laser is running. One such device reported by *Neister* [6.75] used five different dye circulators to obtain rapid tuning between 750 and 950 nm. Such systems must have excellent flushing out of the previous dye, especially when going from shorter wavelengths to longer ones, since absorption by a tiny quantity of the previous dye may quench the lasing.

6.11.1 Injection Locking

An attractive solution for obtaining high gain at high efficiency is the injection-locked oscillator, in which a broad-band high-power oscillator is "seeded" by injecting a carefully controlled signal at the start of the oscillation. This approach can give gain of more than 10^6 with the efficiency of a broad-band oscillator. The seed can come from a low-power FEDL or from another type of laser oscillator. Spectra of only a few megahertz bandwidth have been obtained when injecting with a single-mode laser-pumped cw dye laser. The seed can be injected by any method that does not upset the integrity of the resonator. Best efficiency is obtained by matching the spectral and spatial properties of the injection to the resonator modes, but the matching need not be perfect. Both stable and unstable resonators have been injected. The injection is only signific-ant for a very brief time at the onset of the oscillation. It replaces the noise that the oscillator would otherwise build upon. Thereafter, the output spectrum is fixed until the locking is lost, which may be after nanoseconds or microseconds depending upon the injected intensity, the bandwidth, and how well the spectrum of the injected signal is matched to the peak of the gain curve. Once the seed is injected, the seed oscillator is unimportant, so later feedback that might upset its spectral purity does not matter.

The objective is to maintain the locking during the whole pulse length of the oscillator. If the matching is not good enough, then a frequency-selective element in the resonator centered on the spectrum of the injected signal will help. The author has successfully used angle-tuned solid etalons and birefringent filters for this purpose. They can also be used to suppress a broadening and chirping of the spectrum, generally toward the red, which the author has observed when injection locking with spectra less than 1 GHz wide. A solid etalon with many orders within the gain bandwidth, but straddling the injected spectrum, can be useful to prevent broadening.

The author has used seeds from two oscillators to obtain simultaneous locking in two bands, and also switched between seeds from different oscillators for successive pulses. If a ring oscillator is injected, then the completeness of locking can often be inferred from the ratio of energy flowing in the locked direction to that in the unlocked direction.

Magyar and *Schneider-Muntau* [6.47] demonstrated injection locking in 1972. They injected a 0.4 μs narrow-band oscillator through a mirror shared with an untuned ring oscillator, obtaining a power gain of about 4000 and energy output of about 0.6 J in a width of about 0.01 nm, in a 5 μs pulse. The injected pulse had to be present at the beginning of the growth of the main pulse. *Maeda* et al. [6.48] used the same approach. They injected through a mirror shared with a linear narrow-band oscillator. They obtained 4 J of the output locked to the 4 GHz spectrum of the 15 mJ input from the narrow-band laser, with an overall efficiency of 0.24% into the narrow spectrum. There was an additional approximately 1.7 J not locked. *Boquillon* et al. in 1987 demonstrated 6 MHz bandwidth, 4 mJ, 0.4 μs pulses at 10 Hz [6.50]. *Trehin* et al. [6.171] reported a spectral shift between the locking signal and the output with a ring laser. The output pulses of 1 mJ and 8 MHz bandwidth were shifted toward higher frequencies by 7 MHz from the 8 MHz locking bandwidth, attributed to refractive index changes in the dye during pumping. Such shifts may not matter if the locking signal can be tuned to obtain the desired central wavelength at the output. Review papers on injection locking have been published by *Buczek* et al. [6.172], *Ganiel* et al. [6.173] and *Chow* [6.174]. The review by Ganiel is particularly useful. *Gibson* and *Thomas* demonstrated that mode matching is not required [6.175]. *Flamant* and *Mégie* [6.176] have investigated the parameters that affect the efficiency of injection locking.

There have been reports of other types of lasers with unstable resonators that have been injection-locked through a hole on-axis in one of the mirrors. Some of these may have been in the cross-over regime between multi-pass amplifiers of the type described earlier, and truly injection-locked oscillators. The classes may be differentiated by asking whether the "injection" must be continued throughout the pulse. In true injection locking its presence is immaterial once the oscillation has started. The hole size question is addressed by *Kedmi* and *Treves* [6.177].

6.11.2 Mode Control

For some applications the details of the mode structure must be controlled. Examples include isotope separation and excitation of sodium in the upper atmosphere. The modes of a resonator are separated by $c/2L$ Hz, where c is the velocity of light and L is the length of the resonator. The individual modes may have width about 10% of their separation. Thus, if the laser has high beam quality such that only longitudinal modes are oscillating, then if more than one mode is within the gain bandwidth, only about 10% of the spectral band contains energy. Additionally, laser pulses tend to have ragged filling of the mode structure because of the probabilistic nature of the oscillation. Various techniques are available for filling out the mode structure and for chirping the modes during a pulse. One approach is to modulate the optical length of the resonator during the pulse with a Pockel's cell, applying sinusoidal modulation at the intermode frequency spacing to share energy between the modes. A ramp can be superposed to chirp the modes through more than one mode interval during the pulse. It is convenient to do such modulation within an injection-locked oscillator [6.51]. References on internal modulation of lasers include *Siegman's* book [6.2] and a paper by *Yariv* [6.178].

6.11.3 Amplified Spontaneous Emission

Both the spontaneous emission and the gain that it sees are minimized by working with the lasing gain saturated. Hence, the condition for high lasing efficiency is also the condition for low amplified spontaneous emission (ASE). High-gain amplifiers are more likely to produce significant ASE than are oscillators which inherently run saturated. Injected oscillators suffer from ASE if not completely locked, as discussed above.

6.11.4 Frequency Conversion

Methods for converting the output of a laser to a different frequency include harmonic generation and frequency mixing in non-linear materials and Raman shifting. There is considerable literature covering many different types of lasers, and we will not cover the whole field here other than to say that the methods can generally translate from one type of laser to another. A useful general reference is by *Bass* and *Stitch* [6.179]. When non-linear phenomena are utilized, then high peak intensity is generally needed to obtain useful conversion efficiency. Many materials will be damaged if the intensity needed for useful conversion efficiency is sustained for the pulse durations typical of high power FEDLs. In such cases, mode-locking to obtain a series of shorter, more intense pulses, can improve the situation.

Adhav and *Wallace* discuss KDP isomorphs for frequency doubling of dye lasers [6.180, 181]. *Mutin* and *Boquillon* [6.182] describe mixing two single-mode FEDL beams in a $LiIO_3$ crystal to generate the infrared difference frequency, tunable from 3.5 to 5.9 μm with 6 MHz bandwidth.

Acknowledgments

The author is indebted to a number of individuals and organizations for helpful discussions and information. They include the major flashlamp and dye laser manufacturing organizations; H. R. Aldag, G. S. Janes, D. E. Klimek, D. P. Pacheco, P. S. Rostler and R. E. Schlier who were former colleagues at the Avco Everett Research Laboratory; J. F. Black of Standford University who critically reviewed the text, and D. Goodwin who patiently prepared the manuscript. Any mistakes, however, are those of the author. MIT Lincoln Laboratory has been generous in providing resources. Much of the thinking represented has been developed under funding from DARPA, the US Air Force, MICOM and Avco Everett Research Laboratory (now Avco Research Laboratory, Inc.).

Appendix
A. Correlation of the Dye Models of Chaps. 3 and 6

Dye modeling was developed in Chap. 3 with the emphasis on laser pumping, and in Chap. 6 with emphasis on flashlamp pumping. The approaches were slightly different, as were the nomenclatures. The common result was that ratios of dye parameters are more important than the values of individual parameters.

Table A.1. Parameter relationship between Chaps. 3 and 6

Description	Parameter Chap. 3	Chap. 6
Pump absorption	σ_{01}	σ_{PA}
Lasing absorption	σ_{01}^L	σ_{LA}
Pump emission	—	σ_{PE}
Lasing emission	σ_e	σ_{LE}
Pump singlet absorption	σ_{12}	σ_{P1}
Lasing singlet absorption	σ_{12}^L	σ_{L1}
Pump triplet absorption	σ_T	σ_{PT}
Lasing triplet absorption	σ_T^L	σ_{LT}
Singlet to ground spontaneous	$1/\tau$	γ_{10}
Singlet to triplet spontaneous	k_{ST}	γ_{1T}
Triplet to ground spontaneous	$1/\tau_{TS}$	γ_{T0}
Bulk absorption	—	k
Population density ground	n_0	N_0

Table A.2. Parameters measured by recipes of Chap. 3

Parameter	Definition Chap. 3 notation	Chap. 6 notation
Pump absorption saturation intensity	$I_{PS} = \dfrac{1}{\sigma_{01}\tau}$	$= \dfrac{\gamma_{10} + \gamma_{1T}}{\sigma_{PA}}$
Stimulated emission saturation intensity	$I_S = \dfrac{1}{\sigma_e\tau}$	$= \dfrac{\gamma_{10} + \gamma_{1T}}{\sigma_{LE}}$
Probe laser absorption saturation intensity	$I_{LS} = \dfrac{1}{\sigma_{01}^L\tau}$	$= \dfrac{\gamma_{10} + \gamma_{1T}}{\sigma_{LA}}$
Pump excited state absorption, normalized	$R = \dfrac{\sigma_{12}}{\sigma_{01}}$	$= \dfrac{\sigma_{P1}}{\sigma_{PA}}$
Laser excited state absorption, normalized	$R_e = \dfrac{\sigma_{12}^L}{\sigma_e}$	$= \dfrac{\sigma_{L1}}{\sigma_{LE}}$
Pump absorption cross section	σ_{01}	σ_{PA}
Lasing absorption cross section	σ_{01}^L	σ_{LA}
Singlet to triplet spontaneous emission	k_{ST}	γ_{1T}

Chapter 3 gave recipes for measuring the important parameters. The same measurements will give the parameters used in the modeling of Chap. 6. The parameter equivalents in Table 6A.1 will help in making the transfer.

It will be noted that the Chap. 6 modeling includes the bulk absorption parameter k and the pump emission parameter σ_{PE}. The bulk absorption must be considered when dye degradation is a factor, as is often the case for flashlamp pumping. The emission parameter σ_{PE} is found to be a relatively unimportant factor. It contributes only to the p_{PX} parameter of Chap. 6. This only degrades the efficiency significantly if the dye degradation parameter p_{LB} (which contains k) is large, as can be seen in (6.41).

The recipes in Chap. 3 are for accurately measuring the values of I_{PS}, I_S, I_{LS}, R, R_e, σ_{01} and σ_{01}^L using common laboratory equipment. The more difficult measurement of k_{ST} is also discussed. These parameters are listed and defined in Table 6A.2. Knowledge of these parameters allows most of the parameters of Table 6A.1 to be extracted for the modeling of Chap. 6. The ones still remaining are σ_{PE}, σ_{PT}, σ_{LT}, γ_{TO} and k, which must be obtained or estimated by other means for full modeling.

References

6.1   F.P. Schäfer (ed.): *Dye Lasers, Topics Appl. Phys. Vol.* 1, *3rd edn.* (Springer, Berlin, Heidelberg 1990)
6.2   A.E. Siegman: *Lasers* (University Science Books, Mill Valley, CA 1986)

6.3   S.E. Neister: SPIE, **335**, 36 (1982)
6.4   L.G. Nair: Prog. Quantum Electron. **7**, 153 (1982)
6.5   Y.G. Basov: Instrum. Exp. Tech. **29**, 1239 (1986)
6.6   H.W. Furumoto: SPIE, **609**, 111 (1986)
6.7   R. Sierra: Laser Focus/Electro-Optics, **24** (1988)
6.8   F.J. Duarte, L.W. Hillman (eds.): *Dye Laser Principles With Applications* (Academic, San Diego 1990)
6.9   P.P. Sorokin, J.R. Lankard: IBM J. Res. Dev. **11**, 148 (1967)
6.10 P.P. Sorokin, J.R. Lankard, V.L. Moruzzi, E.C. Hammond: J. Chem. Phys. **48**, 4726 (1968)
6.11 B.B. Snavely, F.P. Schäfer: Phys. Lett. **28A**, 728 (1969)
6.12 J.B. Marling, D.W. Gregg, S.J. Thomas: IEEE J. Quantum Electron. **QE-6**, 570 (1970)
6.13 J.B. Marling, D.W. Gregg, L. Wood: Appl. Phys. Lett. **17**, 527 (1970)
6.14 R. Pappalardo, H. Samelson, A. Lempicki: IEEE J. Quantum Electron. **QE-6**, 716 (1970)
6.15 T.F. Ewanizky, R.H. Wright Jr., H.H. Theissing: Appl. Phys. Lett. **22**, 520 (1973)
6.16 S. Bilt, A. Fisher, U. Ganiel: Appl. Opt. **13**, 335 (1974)
6.17 G. S. Janes, H. R. Aldag, D.P. Pacheco, R.E. Schlier: A long pulse duration good beam quality supersonically "Scanned Beam" dye laser, in Proceedings of the International Conference on Lasers '87, ed. by F.J. Duarte (STS McLean, VA 1988) pp. 356–363
6.18 W. Schmidt, F.P. Schäfer: Phys. Lett. **26A**, 558 (1968)
6.19 C.V. Shank, E.P. Ippen: Ultrashort pulse dye lasers, in [6.1] pp. 139–153
6.20 S. Singh: Appl. Opt. **26**, 66 (1987)
6.21 R.G. Morton, M.E. Mack, I. Itzkan: Appl. Opt. **17**, 3268 (1978)
6.22 J.L. Emmett, A.L. Schawlow: Appl. Phys. Lett. **2**, 204 (1963)
6.23 J. Jethwa, F.P. Schäfer: Appl. Phys. **4**, 299 (1974)
6.24 M.H. Ornstein, V.E. Derr: Appl. Opt. **13**, 2100 (1974)
6.25 H.W. Friedman, R.G. Morton: Appl. Opt. **15**, 1494 (1976)
6.26 J. Jethwa, F.P. Schafer, J. Jasny: IEEE J. Quantum Electron. **QE-14**, 119 (1978)
6.27 C.M. Ferrar: Rev. Sci. Instrum. **40**, 1436 (1969)
6.28 A.J. Gibson: J. Phys. E**5**, 971 (1972)
6.29 P. Anliker, M. Gassmann, H. Weber: Opt. Commun. **5**, 137 (1972)
6.30 M.E. Mack, O.B. Northam, L.G. Crawford: 1976 Annual Meeting Opt. Soc. Am. (1976) p. 1108
6.31 W.W. Morey, W.H. Glenn: IEEE J. Quantum Electron. **12**, 311 (1976)
6.32 E. Thiel, C. Zander, K.H. Drexhage: Opt. Commun. **63**, 171 (1987)
6.33 F.N. Baltakov, B.A. Barikhin, L.V. Sukhanov: JETP Lett. **19**, 174 (1974)
6.34 P. Mazzinghi et al.: IEEE J. Quantum Electron. **QE-17**, 2245 (1981)
6.35 R.G. Morton, V.G. Dragoo: IEEE J. Quantum Electron. **QE-17**, 222 (1981)
6.36 P. Mazzinghi, V. Rivano, P. Burlamacchi: Appl. Opt. **22**, 3335 (1983)
6.37 J.C. Hsia: Private communication (1990)
6.38 W.W. Wladimiroff: Photochemistry and Photobiology **6**, 543 (1967)
6.39 K.L. Matheson, J.M. Thorne: Appl. Phys. Lett. **35**, 314 (1979)
6.40 V. Rivano, P. Mazzinghi: Appl. Phys. B **35**, 71 (1984)
6.41 P.N. Everett et al.: Appl. Opt. **25**, 2142 (1986)
6.42 D.P. Pacheco, H.R. Aldag, P.S. Rostler: The use of spectral conversion to increase the efficiency of flashlamp-pumped dye lasers," in Proceedings of the International Conference on Lasers '88, ed. by R.C. Sze, F.J. Duarte (STS, McLean, VA 1989) pp. 410–419
6.43 M.B. Levin, A.S. Cherkasov: Sov. J. Quantum Electron. **19**, 157 (1989)
6.44 A. Hirth, K. Vollrath, J.Y. Allain: Opt. Commun. **20**, 347 (1977)
6.45 R.M. Schotland: Appl. Opt. **19**, 124 (1980)
6.46 G.M. Gale: Opt. Commun. **7**, 86 (1973)
6.47 G. Magyar, H.-J. Schneider-Muntau: Appl. Phys. Lett. **20**, 406 (1972)
6.48 M. Maeda, O. Uchino, T. Okada, Y. Miyazoe: Jpn. J. Appl. Phys. **14**, 1975 (1975)
6.49 S. Blit, U. Ganiel, D. Treves: Appl. Phys. **12**, 69 (1977)

6.50   J.P. Boquillon, Y. Ouazzany, R. Chaux: J. Appl. Phys. **62**, 23 (1987)

6.51   P.N. Everett: Mode-locking and chirping system for lasers, U.S. Patent No. 4,314,210 (Feb. 2, 1982)

6.52   H.R. Aldag, R.G. Morton, J.A. Woodroffe: High performance dye laser and flow channel therefor, U.S. Patent No. 4,176,324 (Nov. 27, 1979)

6.53   H.R. Aldag, J.B. Marling, C.T. Pike: Flashlamp excited fluid laser amplifier, U.S. Patent No. 4,292,601 (Sept. 29, 1981)

6.54   K. Andringa, C.T. Pike: Laser ring resonator with divergence compensation, U.S. Patent No. 4,203,077 (May 13, 1980)

6.55   K. Avicola, G. Baker: Automatic locking system for an injection-locked laser, U.S. Patent No. 4,264,870 (April 28, 1981)

6.56   P.G. Debarysche, J.L. Munroe: Method and apparatus for sequentially combining pulsed beams of radiation, U.S. Patent No. 3,924,937 (Dec. 9, 1975)

6.57   V.G. Draggoo: Thermally improved dye laser flow channel, U.S. Patent No. 4,296,388 (Oct. 20, 1981)

6.58   J.M. Drake, H.W. Furumoto: Compact reservoir system for dye lasers, U.S. Patent No. 4,364,015 (Dec. 14, 1982)

6.59   P.N. Everett: Laser mode locking, Q-switching and dumping system, U.S. Patent No. 4,375,684 (March 1, 1983)

6.60   H.W. Friedman, J.W. McCloskey, R.M. Patrick: Flashlamp pumped laser device employing fluid material for producing laser beam, U.S. Patent No. 3,992,684 (Nov. 16, 1976)

6.61   I. Itzkan, C.T. Pike: Laser wavelength stabilization, U.S. Patent No. 3,967,211 (June 29, 1976)

6.62   I. Itzkan, C.T. Pike: Apparatus for lengthening laser output pulse duration, U.S. Patent No. 3,914,709 (Oct. 21, 1975)

6.63   R.H. Levy, C.T. Pike: Laser amplifier system, U.S. Patent No. 3,944,947 (Mar. 16, 1976)

6.64   M.E. Mack, D.B. Northam: Stabilized repetitively pulsed flashlamps, U.S. Patent No. 4,074,208 (Feb. 14, 1978)

6.65   M.E. Mack, J.A. Woodroffe: Flow channel for fluid medium laser, U.S. Patent No. 4,170,762 (Oct. 9, 1979)

6.66   G.L. McAllister: Controlled frequency high power laser, U.S. Patent No. 4,181,898 (Jan. 1, 1980)

6.67   G.L. McAllister, V.G. Draggoo: Multiwavelength dye laser, U.S. Patent No. 4,293,827 (Oct. 6, 1981)

6.68   G.L. McAllister: Method and apparatus for improving flashlamp performance, U.S. Patent No. 4,469,991 (Sept. 4, 1984)

6.69   R.G. Morton: High pulse repetition rate coaxial flashlamp, U.S. Patent No. 4,325,006 (April 13, 1982)

6.70   R.G. Morton: Fluid laser flow channel liner, U.S. Patent No. 4,178,565 (Dec. 11, 1979)

6.71   R.G. Morton: Unstable waveguide laser resonator, U.S. Patent No. 4,423,511 (Dec. 27, 1983)

6.72   C.T. Pike: Improved apparatus for lengthening laser output pulse duration, U.S. Patent No. 3,902,130 (Aug. 26, 1975)

6.73   D.P. Pacheco, H.R. Aldag, D.E. Klimek, P.S. Rostler, R. Scheps, J.F. Myers: A high-average-power blue-green laser for underwater communications, in Proceedings of the International Conference on Lasers '89, ed. by D.G. Harris, T.M. Shay (STS, McLean, VA, 1990) pp. 376–382

6.74   P.N. Everett: Performance and modeling of a flashlamp-pumped dye laser with aqueous acetamide as a solvent, in Proceedings of the International Conference on Lasers '89, ed. by D.G. Harris, T.M. Shay (STS, McLean, VA 1990) pp. 393–402

6.75   S.E. Neister: Private communication (1990)

6.76   I.S. Marshak: Sov. Phys. Uspekhi **5**, 478 (1962)

6.77   I.S. Marshak: Appl. Opt. **2**, 793 (1963)

6.78   J.H. Goncz: ISA Transactions **5**, 28 (1966)

6.79   H.W. Furumoto, H.L. Ceccon: Appl. Opt. **8**, 1613 (1969)

6.80   "Flashlamp Applications Manual," in EG & G Electro-Optics Bulletin (1982)

6.81   "An Introduction to Flashlamps, Technical Bulletin 1," in ILC Technology (1986)
6.82   "An Overview of Flashlamps and cw Arc Lamps, Technical Bulletin 3," in ILC Technology (1986)
6.83   L.A. Godfrey, R.S. Meltzer, J.E. Rives: Appl. Opt. **18**, 602 (1979)
6.84   A. Hirth, T. Lasser, R. Meyer, K. Schetter: Opt. Commun. **34**, 223 (1980)
6.85   A. Buck, R. Erickson, F. Barnes: J. Appl. Phys. **34**, 2115 (1963)
6.86   R.H. Dishington, W.R. Hook, R.P. Hilberg: Appl. Opt. **13**, 2300 (1974)
6.87   A. Marotta, C.A. Arguello: J. Phys. E. **9**, 478 (1976)
6.88   T.K. Yee, B. Fan, T.K. Gustafson: Appl. Opt. **18**, 1131 (1979)
6.89   Y.G. Basov, T.I. Mikhalina, V.G. Nikiforov, A.I. Sopin: J. Appl. Spectrosc. **32**, 329 (1980)
6.90   J.P. Markiewicz, J.L. Emmett: IEEE J. Quantum Electron. **QE-2**, 707 (1966)
6.91   D.E. Perlman: Rev. Sci. Instrum. **37**, 340 (1966)
6.92   T.F. Ewanizky, R.H. Wright Jr.: Appl. Opt. **12**, 120 (1973)
6.93   J.H. Goncz, P.B. Newell: J. Opt. Soc. Am. **56**, 87 (1966)
6.94   J.L. Emmett, A.L. Schawlow, E.H. Weinberg: J. Appl. Phys. **35**, 2601 (1964)
6.95   K. Gunther: Opt. Spectrosc. **34**, 620 (1973)
6.96   V.E. Gavrilov, O.B. Danilov, A.P. Zhevlakov, S.A. Tul'Skil: Sov. J. Opt. Technol. **49**, 403 (1982)
6.97   B.P. Giterman et al.: Sov. Phys. Tech. Phys. **27**, 1218 (1982)
6.98   P.A. Lovoi: SPIE **138**, 2 (1978)
6.99   J.H. Kelley, D.C. Brown, K. Teegarden: Appl. Opt. **19**, 3817 (1980)
6.100   W.L. Wolfe, G.J. Zissis (eds.): *The Infrared Handbook* (Office of Naval Research, Washington, DC 1978)
6.101   D.P. Pacheco: Private communication (1989)
6.102   T. Efthymiopoulos, B.K. Garside: Appl. Opt. **16**, 70 (1977)
6.103   T. Okada, K. Fujiwara, M. Maeda, Y. Miyazoe: Appl. Phys. **15**, 191 (1978)
6.104   M.E. Mack: Appl. Opt. **13**, 46 (1974)
6.105   J.F. Holzrichter, J.L. Emmett: Appl. Opt. **8**, 1459 (1969)
6.106   V.A. Alekseev et al.: Sov. J. Quantum Electron. **1**, 643 (1972)
6.107   M. Held: 8th Conference on High Speed Photography, Stockholm (1968)
6.108   F.P. Schäfer: Conference on Nonlinear Optics, Belfast (1969)
6.109   S.A. Baranov, I.V. Kolpakova, M.Y. Kononova, A.A. Mak, O.A. Motovilov: Sov. J. Quantum Electron. **8**, 102 (1978)
6.110   R. Phillips: *Sources and Applications of Ultraviolet Radiation* (Academic, London 1983)
6.111   EG & G Data Sheets (March 1988)
6.112   Y.A. Kalinin, A.A. Mak: Sov. J. Opt. Tech. **37**, 129 (1970)
6.113   M.M. Weiner: J. Opt. Soc. Am. **54**, 1109 (1964)
6.114   W.T. Welford, R. Winston: *The Optics of Nonimaging Concentrators, Light and Solar Energy* (Academic, New York 1978)
6.115   P. Burlamacchi, R. Pratesi, R. Salimbeni: Opt. Commun. **11**, 109 (1974)
6.116   P. Burlamacchi, R. Pratesi, U. Vanni: Rev. Sci. Instrum. **46**, 281 (1975)
6.117   P. Burlamacchi, R. Pratesi: Appl. Phys. Lett. **28**, 124 (1976)
6.118   P. Burlamacchi, R. Pratesi, U. Vanni: Appl. Opt. **15**, 2684 (1976)
6.119   B.B. Snavely: Proc. IEEE (1969)
6.120   B.B. Snavely: Continuous-wave dye lasers I, in Ref. [6.1] pp. 91–120
6.121   A. Einstein: Physik. Zeitschr. **18**, 121 (1917)
6.122   U. Ganiel, A. Hardy, G. Neumann, D. Treves: IEEE J. Quantum Electron. **QE-11**, 881 (1975)
6.123   R.S. Hargrove, T. Kan: IEEE J. Quantum Electron. **QE-16**, 1108 (1980)
6.124   P.N. Everett: Limits on efficiency of optically pumped dye lasers, in Proceedings of the International Conference on Lasers '89, ed. by D.G. Harris, T.M. Shay (STS, McLean, VA 1990) pp. 383–392
6.125   W.W. Morey: Copper vapor laser pumped dye laser, in Proceedings of the International Conference on Lasers '79, ed. by V.J. Corcoran (STS, McLean, VA 1980) pp. 365–373
6.126   O. Uchino, T. Mizunami, M. Maeda, Y. Miyazoe: Appl. Phys. **19**, 35 (1979)

6.127 F. Bos: Appl. Opt. **20**, 1886 (1981)

6.128 P.R. Hammond: IEEE J. Quantum Electron. **QE-16**, 1157 (1980)

6.129 D.E. Klimek: Private communication (1984)

6.130 R.H. Khipe, A.N. Fletcher: J. Photochem. **23**, 117 (1983)

6.131 G. Jones II.: Photochemistry and photophysics of laser dyes, Office of Naval Research, Tech. Rept. No. 8 (October 31, 1983)

6.132 G. Jones II, S.F.Griffin, C. Choi, W.R. Bergmark: Electron donor-acceptor quenching and photo-induced electron transfer for coumarin dyes, Office of Naval Research, Tech. Rept. No. 3 (October 31, 1983)

6.133 G. Jones II., W.R. Bergmark: Photodegradation of coumarin laser dyes, Office of Naval Research, Tech. Rept. No. 4 (October 31, 1983)

6.134 G. Jones II., W.R. Jackson, C. Choi: Solvent effects on emission yield and lifetime for coumarin laser dyes. Requirements for a rotatory decoy mechanism, Office of Naval Research, Tech. Rept. No. 5 (October 31, 1983)

6.135 G. Jones II., W.R. Jackson, S. Kanoktanoporn, W.R. Bergmark: Photophysical and photochemical properties of coumarin laser dyes in amphiphilic media, Office of Naval Research, Tech. Rept. No. 6 (October 31, 1983)

6.136 G. Jones II., W.R. Jackson, S. Kanoktanoporn, W.R. Bergmark: Products of Photodegradation for coumarin laser dyes, Office of Naval Research, Tech. Rept. No. 7 (October 31, 1983)

6.137 P.N. Everett: 300 Watt dye laser for field experimental site, in Proceedings of the International Conference on Lasers '88, ed. by R.C. Sze, F.J. Duarte (STS, McLean, VA 1989) pp. 404–409

6.138 P.G. Weber: IEEE J. Quantum Electron. **QE-19**, 1200 (1983)

6.139 O.G. Peterson, B.B. Snavely: Bull. Am. Phys. Soc. **13**, 397 (1968)

6.140 C.E. Moeller, C.E. Verber, A.H. Adelman: Appl. Phys. Lett. **18**, 278 (1971)

6.141 P. Burlamacchi, D. Cutter: Opt. Commun. **22**, 283 (1977)

6.142 T.G. Pavlopoulos: Opt. Commun. **24**, 170 (1978)

6.143 A.N. Fletcher: Appl. Phys. B **31**, 19 (1983)

6.144 P. Lacovara, L. Esterowitz, R. Allen: Opt. Lett. **10**, 273 (1985)

6.145 P. Lacovara: Private communication (1987)

6.146 K.J. Nygaard: J. Opt. Soc. Am. **55**, 944 (1965)

6.147 N. Kritianpoller, R.A. Knapp: Appl. Opt. **3**, 915 (1964)

6.148 "Parker O-ring Sizes and Compound Reference Guide, Parker Hannifin Corp., O-ring Division

6.149 'Chemical Resistance Guide", Met Pro Corp., Sethco Division

6.150 Cole-Parmer Catalog (1987–1988)

6.151 F.J. Duarte, J.J. Ehrlich, S.P. Patterson, S.D. Russel, J.E. Adams: Appl. Opt. **27**, 843 (1988)

6.152 E.A. Gavronskaya, A.V. Groznyi, D.I. Staselko, V.L. Strigun: Opt. Spectrosc. **42**, 213 (1977)

6.153 A.V. Aristov, D.A. Kozlovskii, D.I. Staselko, V.L. Strigun: Opt. Spectrosc. (USSR) **45**, 683 (1978)

6.154 W.H. Steier: Unstable resonators, in Laser Handbook, ed. by M.L. Stitch (North-Holland, Amsterdam 1979) pp. 3–39

6.155 E.B. Brown: *Modern Optics* (Krieger, Malabar, FL 1974) pp. 225 & 229

6.156 M.I. Skolnik: *Radar Handbook* (McGraw Hill, New York 1970)

6.157 J. Strong: *Concepts of Classical Optics* (Freeman, San Fransisco 1958) p. 54

6.158 A. Yariv: *Quantum Electronics* (Wiley, New York 1975)

6.159 A.E. Siegman: Proc. IEEE **53** (1965)

6.160 L.V. Koval'chuk, N.A. Sventisitskaya: Sov. J. Quantum Electron. **2**, 450 (1973)

6.161 S.L. Chao, A.D. Schnurr: Appl. Opt. **23**, 2115 (1984)

6.162 H. Kogelnik, T. Li: Appl. Opt. **5**, 1550 (1980)

6.163 A.A. Isaev, M.A. Kazaryan, G.G. Petrash, S.G. Rautian: Sov. J. Quantum Electron. **4**, 761 (1974)

6.164 J.J. Degnan: Appl. Phys. **11**, (1976)

6.165 R.P. Feynman, R.B. Leighton, M. Sands: *The Feynman Lectures on Physics* (Addison-Wesley, Reading, MA 1963)
6.166 F.A. Jenkins, H.E. White: *Fundamentals of Optics* (McGraw-Hill, New York 1976)
6.167 P.P. Clark, J.W. Howard, E.R. Freniere: Appl. Opt. **23**, 353 (1984)
6.168 V.G. Draggoo, R.G. Morton, R.H. Sawicki, H.D. Bissinger: SPIE **622**, 186 (1986)
6.169 C.V. Shank, E.P. Ippen: Appl. Phys. Lett. **24**, 122 (1974)
6.170 W. Sibbett, J.R. Taylor: IEEE J. Quantum Electron. **QE-18**, 1994 (1982)
6.171 F. Trehin, B. Biraben, B. Cagnac, G. Grynberg: Opt. Commun. **31**, 76 (1979)
6.172 C.J. Buczek, R.J. Freiberg, M.L. Skolnick: Proc. IEEE **61**, 1411 (1973)
6.173 U. Ganiel, A. Hardy, D. Treves: IEEE J. Quantum Electron. **QE-12**, 704 (1976)
6.174 W.W. Chow: IEEE J. Quantum Electron. QE-19, 243 (1983)
6.175 A.J. Gibson, L. Thomas: J. Phys. D11, L59 (1978)
6.176 P. Flamant, G. Mégie: IEEE J. Quantum Electron. **QE-16**, 653 (1980)
6.177 J. Kedmi, D. Treves: Appl. Opt. **20**, 2108 (1981)
6.178 A. Yariv: J. Appl. Phys. **36**, 388 (1964)
6.179 M. Bass, M.L. Stitch (eds.) *Laser Handbook* (North-Holland, Amsterdam 1985)
6.180 R.S. Adhav, R.W. Wallace: IEEE J. Quantum Electron. **QE-9**, 855 (1973)
6.181 R.S. Adhav: Materials for optical harmonic generation, Laser Focus (1983)
6.182 P. Mutin, J.P. Boquillon: Appl. Phys. B **48**, 411 (1989)

Subject Index

Springer Series in Optical Sciences

Editorial Board: A. L. Schawlow K. Shimoda A. E. Siegman T. Tamir